The giant globular cluster Omega Centauri This VST image of Omega Centauri may be the best portrait of the globular star cluster ever made. VST's OmegaCAM can encompass even the faint outer regions of this spectacular object.

EUROPE
TO THE
STARS

ESO'S FIRST 50 YEARS OF EXPLORING THE SOUTHERN SKY

Europe to the Stars —
ESO's First 50 Years of Exploring the Southern Sky

The Authors
Govert Schilling & Lars Lindberg Christensen

Design & Layout
ESO education and Public Outreach Department
André Roquette & Francesco Rossetto

Library of Congress Card No.
Applied for

British Library Cataloguing-in-Publication Data
A catalogue record for this book is available from the British Library.

Bibliographic information published by the Deutsche Nationalbibliothek
The Deutsche Nationalbibliothek lists this publication in the Deutsche Nationalbibliografie; detailed bibliographic data are available on the Internet at <http://dnb.d-nb.de>.

Printed and binding
Himmer AG, Augsburg, Germany
Printed on acid-free paper
ISBN: 978-3-527-41192-4

Cover and back
The VLT
This photograph taken by ESO Photo Ambassador Babak Tafreshi, captures the ESO Very Large Telescope (VLT) against a beautiful twilight sky on Cerro Paranal. A VLT enclosure stands out in the picture as the telescope is readied for a night studying the Universe. The VLT is the world's most powerful optical telescope, consisting of four Unit Telescopes with primary mirrors of 8.2-metre diameter and four movable 1.8-metre Auxiliary Telescopes, which can be seen on the back of the cover. Over the past 13 years, the VLT has had a huge impact on observational astronomy. With the advent of the VLT, the European astronomical community has experienced a new age of discoveries, most notably, the tracking of the stars orbiting the Milky Way's central black hole and the first image of an extrasolar planet.

Back, top
Heavenly wonders
A selection of spectacular images made with ESO's telescopes.

To ESO's unsung heroes

ESO — Reaching New Heights in Astronomy

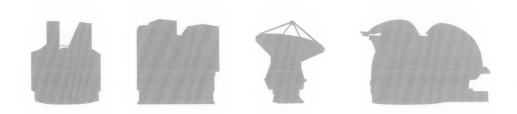

Foreword

The signing of the ESO Convention in 1962 and the creation of ESO was the culmination of the dream of leading astronomers from five European countries, Belgium, France, Germany, the Netherlands and Sweden: a joint European observatory to be built in the southern hemisphere to give astronomers from Europe access to the magnificent and rich southern sky by means of a large telescope. The dream resulted in the creation of the La Silla Observatory near La Serena in Chile and eventually led to the construction and operation of a fleet of telescopes, with the 3.6-metre telescope as flagship. As Italy and Switzerland joined ESO in 1982 the construction of the New Technology Telescope, with pioneering advances in active optics, became possible, preparing the way for the next step: the construction of the Very Large Telescope. The VLT made adaptive optics and interferometry available to a wide community.

The decision to build a fully integrated VLT system, consisting of four 8.2-metre telescopes and providing a dozen foci for a carefully thought-out complement of instruments opened a new era in ESO's history. The combination of a long-term, adequately-funded instrument and technology development plan, with an approach where instruments are built in collaboration with institutions in the Member States, and with in-kind contributions in labour compensated by guaranteed observing time, has created the most advanced ground-based optical observatory in the world.

Today, in 2012, the original hopes of the five founding members have not only become reality but ESO has fully taken up the challenge of its mission to design, build and operate the most powerful ground-based observing facilities on the planet. On the Chajnantor Plateau in Northern Chile, together with North American and East Asian partners, ESO is developing the biggest ground-based astronomical project in existence, the Atacama Large Millimeter/submillimeter Array (ALMA). And ESO is starting to build the world's biggest eye on the sky, the European Extremely Large Telescope.

In 2012, our 50th anniversary year, we are ready to enter a new era, one that not even the initial bold dreams of ESO's founding members could have anticipated. It is undoubtedly a most exciting time that we live in. It is a pleasure to thank everyone involved in making the ESO dream come true: to the ESO staff for their professionalism, ingenuity and passion, to Council and Committee members and the former Directors General for leading the observatory to new heights in astronomy. And to the public, educators and media who on a daily basis take part in ESO's discoveries.

The year 2012 is also a time to congratulate all our Member States. The five founding members have been joined by Denmark (1967), Switzerland (1982), Italy (1982), Portugal (2001), the United Kingdom (2002), Finland (2004), Spain (2007), the Czech Republic (2007), Austria (2009), and Brazil, who will become the 15th, as well as the first non-European, Member State after parliamentary ratification of the Accession Agreement signed in December 2010. The Member States have adhered to ESO's courageous plans to lead ground-based astronomy, and offer us constant support and top-level people. Together these 15 countries contain approximately 30% of the world's astronomers, and by now ESO is the most productive ground-based observatory in the world supplying data for more than 750 scientific papers per year.

The scientific community is to be congratulated for keeping astronomy at the forefront of scientific research, as well as our supporters and international partners for believing in our ambitious projects. ESO owes its success in a large part to these collaborations!

Welcome to the world of ESO!

Tim de Zeeuw
ESO Director General
Garching, June 2012

The VLT at work
The Very Large Telescope with galaxies Messier 31 (left middle) and Messier 33 (top left) as backdrop.

Preface

As part of the celebration of its 50th anniversary, this book paints a portrait of the European Southern Observatory in accessible text and stunning images. Although it presents some historic detail, it is not meant to be a formal history of ESO. Rather, we have focussed on ESO's achievements — on the magnificent telescopes and instruments that enthral anyone visiting them, and on the scientific breakthroughs that they regularly produce. Our visual journey tries to give a feel for ESO as a scientific organisation and to present a cross-section of the many parts that combine in making it the successful endeavour that it is.

In a book like this, it is impossible to cover the topic completely, and also to describe the many wonderful scientific and engineering breakthroughs made outside the ESO-sphere. It is equally not possible to give proper credit to the many people who deserve it. The limited space has only allowed us to introduce the Directors General who have been at the helm of ESO. Anyone interested in a much more comprehensive history of ESO should read *Jewel on the Mountaintop* by Claus Madsen (see p. 254), published together with this book, as the other half of what can really be seen as a complementary set. We have however made space to show some of the many unsung heroes who usually remain invisible, but who represent the cement that holds the organisation together.

The last part of the book provides credits for the many images, further literature about ESO and also the first ever full overview of ESO's telescopes. These facts were collected as part of the anniversary efforts with the help of many ESO employees and especially volunteer Philip Corneille (FBIS/VVS) from Belgium. We are grateful for this assistance. We would also like to thank several individuals for thoughtful comments and corrections: Tim de Zeeuw, Bruno Leibundgut, Gero Rupprecht, Olivier Hainaut, Adam Hadhazy, Mathieu Isidro, Richard Hook, Douglas Pierce-Price, Sally Lowenstein, Mark Casali and Anne Rhodes.

A big thank you also goes to Andre Roquette, Francesco Rossetto, Jutta Boxheimer, Mafalda Martins and Kristine Omandap from ESO's education and Public Outreach Department for the wonderful design of the book, as well as to Mathieu Isidro for the hard work of updating ESO's timeline. Many other individuals from this department, including Hännes Heyer, have also made a major effort, and have helped to put together a treasure trove of more than 7000 photos online over the past four years. We are especially indebted to the world-class photographers who have provided material for this book, most notably ESO Photo Ambassadors Babak Tafreshi, Christoph Malin, José Francisco Salgado, Serge Brunier, Stéphane Guisard, Gerhard Hüdepohl, Gianluca Lombardi, Yuri Beletsky and Gabriel Brammer. These photographers have also in part delivered stunning time-lapse footage for the 60-minute movie that accompanies this book. Max Alexander took the wonderful portraits of unsung ESO heroes for the book, for which we are thankful.

Govert Schilling & Lars Lindberg Christensen
Amersfoort, the Netherlands & Garching bei München, Germany, June 2012

The Chilean night sky at ALMA
This image shows the night sky seen from the Atacama Desert. This photograph was taken from the site of the ALMA cultural heritage museum.

1

Setting the Scene

Today's astronomers who venture south of the equator to stargaze are not the first. Some parts of the Universe can only be observed from the southern hemisphere. Ever since seafarers and explorers first marvelled at the splendour of the Milky Way and the Magellanic Clouds, scientists have been lured to southern latitudes, where unknown constellations held the promise of great discoveries.

The southern sky at the coast of the Chilean Atacama Desert
Because of the humidity over the cold Pacific Ocean, clouds often cover the coast of the Atacama Desert only 12 kilometres from ESO's Paranal Observatory. The cold ocean keeps the so-called inversion layer very low, and the atmosphere above the clouds exceptionally dry and clear.

"Where's Oort?"

The young Dutch astronomer Gart Westerhout hadn't seen his Leiden Observatory professor for at least fifteen minutes. His colleague, Fjeda Walraven, also had no clue as to the whereabouts of the famous scientist. And yes, that was worrisome, for Westerhout and Walraven were carrying out test observations in a pitch-dark field at Hartebeespoort in South Africa, with wild baboons and other animals wandering around the camp. And now, Jan Oort had disappeared, on his very first visit to the southern hemisphere.

Decades later, Westerhout vividly recalled his 1952 experience. *"We found you on the other side of a small hill,"* he wrote on the occasion of Oort's 80th birthday in 1980, *"flat on your back in the wet grass, risking pneumonia, with the centre of the Milky Way in the zenith. You could not be convinced to get up, and you shooed us off! I have never forgotten the impression this event made on me. Here was the man who was the first to unravel the structure of the Galactic System, twenty five years earlier, and who now saw it for the first time, as a natural phenomenon, of which man is a part."*

Jan Oort, who would later become the chief initiator of the European Southern Observatory (ESO), had never before witnessed such an impressive sight. Myriads of tiny, twinkling stars; shimmering clouds of nebulous gas, and wispy streaks of dark dust — all stretched out in a luminous band across the velvet-black sky. Nowhere in Europe, let alone in his small and densely populated home country, could the Milky Way be observed in such magnificent splendour. You just *had* to go south of the equator.

We have evolved on a small, rocky planet, orbiting an inconspicuous star on the outskirts of an undistinguished spiral galaxy. From the north pole of this tiny, rotating sphere, only half the Universe can be seen, and wherever you are located in the northern hemisphere, there's always a sizeable chunk of sky that remains invisible at all times — and this missing chunk of sky contains some of the most spectacular celestial sights.

Jan Oort
The chief initiator of the European Southern Observatory.

Ancient cultures in southern Africa, Latin America and Australia knew all about the beauty of the southern sky. They developed myths and legends concerning the Milky Way and its blazing centre; tales of the hazy patches of light that we now call the Magellanic Clouds, and the many bright stars that pepper the southern skies. But in the Near East and in Europe much of this cosmic scenery never rose above the horizon. Just as maps of the ancient world contained uncharted regions with ominous texts like *"Here be dragons"*, maps of the sky also had blank, unexplored spots.

Remarkably, the famous constellation of the Southern Cross *was* known in Europe. In the time of the ancient Greeks, it just rose above the southern horizon in Athens every April, its stars being considered part of the constellation of Centaurus. But because of an extremely slow change in the cosmic orientation of Earth's axis, first discovered by Hipparchus of Nicaea, the Cross has now disappeared from view for anyone living north of Cairo. It was rediscovered by Portuguese seafarers, and eventually ended up on the national flags of Australia, Brazil, New Zealand, Papua New Guinea, and Samoa.

As for the Magellanic Clouds, the larger of the two was first mentioned by Persian astronomer Abu al-Husan Abd al-Rahman ibn Omar al-Sufi al-Razi — usually referred to simply as al-Sufi — in his 964 AD treatise, *The Book of Fixed Stars*. The cosmic cloud was just barely visible from the southernmost point of Arabia. But here again, knowledge of the celestial fuzz was lost, only to be regained after European explorers set sail for distant shores and marvelled at the new vistas above their heads.

Named after Ferdinand Magellan, who was the first to circumnavigate the world in 1519–1522, the two clouds are now known to be satellite galaxies of the Milky Way. Compared to our home galaxy, the clouds are smaller, irregularly shaped, and have a relatively higher abundance of interstellar gas to spawn new stars. Nevertheless, they are galaxies in their own right, and the nearest ones that astronomers can study in detail. This scientific privilege, however, is only provided when you can set up your telescope equipment south of the terrestrial equator — the clouds are invisible from North America, Asia and Europe.

Pieter Platevoet, a Flemish astronomer, cartographer and clergyman who moved to Amsterdam in 1585, was not able to travel to the southern hemisphere himself. Instead, he taught Dutch seafarers Pieter Dirkszoon Keyser and Frederik de Houtman how to use a cross staff and an astrolabe — simple instruments for measuring stellar positions. Would Keyser and de Houtman be so kind as to map the unknown southern sky during their pioneering spice expedition to the East Indies? If so, Platevoet (better known by his Latin name Petrus Plancius) would finally be able to fill in the blank areas on the charts of the heavens.

The Small Magellanic Cloud over the Chilean landscape
Snow-covered trees under a magnificent night sky, at Torres del Paine National Park, southern Patagonia. Chile's magnificent desert skies are renowned for their clarity, but stargazing can also be impressive in the southern part of this long country. The two brightest stars in the prominent Milky Way band are Alpha (above) and Beta (below) Centauri.

Treasures of the southern sky

Many celestial treasures can best be observed from the southern hemisphere. For instance, the famous Trifid and Lagoon Nebulae in the constellation of Sagittarius (The Archer) never rise high above the horizon as seen from Europe. The same is true for the Rho Ophiuchi star-forming region and the globular cluster Messier 4 in Scorpius.

Even further south are the Carina Nebula and the R Coronae Borealis region — two other stellar nurseries. Omega Centauri and 47 Tucanae are the two most impressive globular clusters in the sky, and the Jewel Box in the Southern Cross is a serious competitor to the Pleiades in the beauty contest for the most impressive open cluster in the sky.

And while the centre of the Milky Way galaxy (middle, opposite page) and the two Magellanic Clouds — nearby satellites of the Milky Way — will never cease to impress, more distant spirals and ellipticals also claim astronomers' attention, like the beautiful spiral NGC 1232 in the constellation of Eridanus, the members of the Sculptor and Fornax clusters, and, last but not least, the distinctive active galaxy Centaurus A.

A 340-million-pixel Paranal starscape
This spectacular 34 by 20 degree-wide image shows one of the most interesting areas of cosmic real estate in the southern sky. Noteworthy objects are the centre of the Milky Way (in the dust lane left) as well as the Trifid, and Omega Nebulae (left) in the constellation of Sagittarius (The Archer). In Scorpius (right) we see the Rho Ophiuchi star-forming region and globular cluster Messier 4. The image was composed from 1200 individual photos taken by ESO engineer Stéphane Guisard.

The expedition was a disaster. Of the 248 men who left the Dutch island of Texel on 2 April 1595, only 81 survived the two and a half year trip. Keyser died on Sumatra, but de Houtman returned to Amsterdam, carrying with him the sky positions of over 130 stars around the south celestial pole. Plancius grouped these into twelve new constellations, including the Bird of Paradise, the Toucan, the Goldfish, the Peacock and the Indian. Within a few years, the new southern constellations were firmly established by cartographers Jodocus Hondius and Willem Janszoon Blaeu, who depicted them on their celestial globes, and by the German astronomer, Johann Bayer, who adopted them in his famous 1603 star atlas *Uranometria*.

Abbé Nicolas Louis de Lacaille greatly extended the work begun by Plancius. In the middle of the 18th century, some 150 years after the invention of the telescope, this French astronomer sailed to the Cape of Good Hope, where he catalogued 10 000 stars in the southern sky. Lacaille also introduced thirteen new constellations, which he named after scientific instruments and equipment, like the Telescope (of course!), the Microscope, the Pendulum Clock, and the Oven. One constellation was called Mons Mensa (Table Mountain), for the location of Lacaille's observatory.

Lacaille was also one of the first to study the Magellanic Clouds in some detail. He noted that one particular star in the Large Magellanic Cloud, originally listed as 30 Doradus, was actually a small nebula. Only later did it become clear that 30 Doradus, also known as the Tarantula Nebula for its spidery filaments, is by far the largest star-forming region in the local Universe, measuring hundreds of light-years across. Hidden in its core is an extremely compact cluster of sizzling newborn stars, no more than two million years old. One of those, R136a1, is actually the most massive and most luminous star known, weighing in at 265 solar masses and pouring out almost nine million times as much energy as the Sun.

John Herschel, son of the legendary William Herschel who discovered the planet Uranus in 1781, knew nothing about the true nature of the Magellanic Clouds when he travelled to the Cape in November 1833. Thirteen years earlier, English astronomers had established the Royal Observatory at the Cape of Good Hope. But Herschel brought his own 46-centimetre telescope, set up a private observatory at the Feldhausen estate in Wynberg, and spent five years cataloguing double stars, star clusters and nebulae, to extend the work his father had begun in the northern hemisphere (actually, the very first southern hemisphere observatory was Georg Marcgrave's 1639 rooftop observatory in Recife, Brazil).

The advent of photography revolutionised the exploration of the night sky. Scottish astronomer David Gill, who was appointed Her Majesty's Astronomer at the Cape

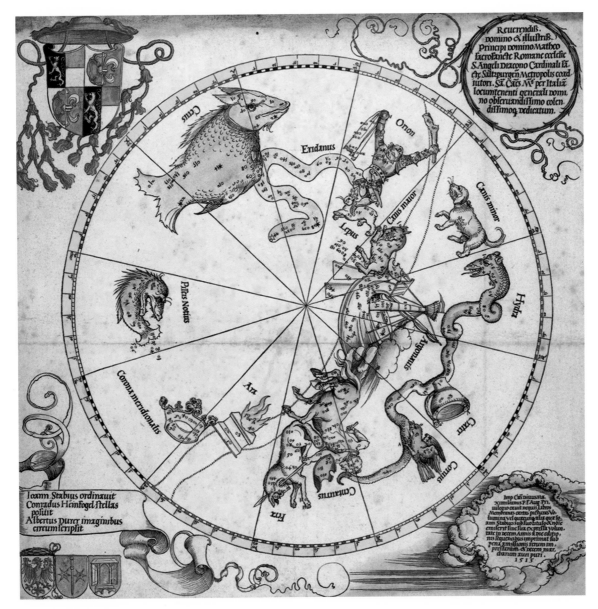

Terra Incognita of the heavens

For many centuries, maps of the southern sky, like this 1515 star chart by Albrecht Dürer, showed extensive blank areas — the *Terra Incognita* of the heavens.

Observatory in 1879, set out to photograph the entire southern sky, using the 61-centimetre McClean telescope. Measuring the positions — and, in many cases, the slow progress across the sky — of 454 875 stars on Gill's glass negatives was a tremendously tedious task, carried out over a period of four years by the Dutch astronomer Jacobus Kapteyn, who would later be Jan Oort's teacher. The resulting *Cape Photographic Durchmusterung* was the first star catalogue based on astrophotography.

By now, scientists all over the world were very much aware that the scarcely populated, arid scrublands of South Africa represented everything an astronomer could ever dream of. Dark skies, cloudless nights, perfect *seeing* — a measure of the lack of atmospheric turbulence — and of course a splendid view of the Magellanic Clouds, the centre of the Milky Way with its countless star clusters and nebulae, and the stars and galaxies of the southern constellations. Staying at home in the northern hemisphere would be like standing on a mountaintop and only enjoying the view in one direction, without ever turning around to admire the much more impressive scenery behind your back.

The Royal Observatory at the Cape of Good Hope
The British were the first to construct a permanent astronomical outpost in the southern hemisphere. The Royal Observatory at the Cape of Good Hope was founded in 1820.

New observatories had already been erected in South Africa. While the Natal Observatory in Durban lasted only from 1882 to 1911, the Transvaal Observatory (later renamed the Union Observatory and the Republic Observatory), was established in 1903 in Johannesburg, and remained operational until the early 1970s (just like Radcliffe Observatory in Pretoria, which was constructed in 1939). Also, a number of American and European universities decided to build their own "southern station" in South Africa. "Venturing south" was the astronomical mantra throughout the 1920s, it seemed.

The Yale–Columbia Southern Station in Johannesburg was the first of these in 1925, sporting a 66-centimetre refractor. Two years later, the University of Michigan constructed the Lamont-Hussey Observatory near Bloemfontein, with a similar-sized telescope. And around the same time, the Harvard College Observatory moved its Boyden Station with its 61-centimetre Bruce telescope from Arequipa, Peru, to Bloemfontein, because of the better weather conditions there.

So what about Jan Oort and his Milky Way encounter in Hartebeespoort? Well, after studying in Groningen with Kapteyn, Oort had worked at Yale for two years before accepting a position at the Leiden Observatory in 1924, so he was well aware of the potential of an astronomical foothold in South Africa. But the Dutch played it a bit differently: In 1923, they had struck a deal with the Union

Observatory, providing astronomers from both institutions with access to each other's instruments. Not surprisingly, given the generally poor weather in the Netherlands, many more Dutch astronomers travelled south to observe in Johannesburg than South Africans came north.

In 1929, Leiden sent its own Rockefeller twin 40-centimetre telescope plus a permanent staff member to the Union Observatory. By the early 1950s, however, the increasing light pollution from Johannesburg became too severe for serious observations, and Dutch astronomers started to scout for a better site. This is why, in 1952, Gart Westerhout and Fjeda Walraven ended up with their test equipment in Hartebeespoort, west of Pretoria. Two years later, the Leiden Southern Station would start operations there, and in 1957, the 90-centimetre Dutch Flux Collector would become one of the largest telescopes in South Africa.

So where was Oort?

Physically, the 52-year old astronomer was lying flat on his back in the wet grass, captivated by the incredible view of the Milky Way. But in his mind, he may have been decades away, in a distant era where European countries would join forces and work together in exploring the southern sky. A few months later, back in the Netherlands, Oort opened discussions with fellow astronomers that would eventually lead to the birth of the European Southern Observatory.

Director General Otto Heckmann

Director General
Otto Heckmann
Painting of Otto
Heckmann, ESO Direc-
tor General between
1962–1969.

> **Name:** Otto Heckmann
> **Year of Birth:** 1901
> **Nationality:** German
> **Period as Director General:** 1962–1969

Heckmann died on 13 May 1983. This "interview" is based on his 1976 book *Sterne, Kosmos, Weltmodelle* (see Further Reading on page 252 below).

What was the greatest challenge during your ESO career?
The greatest challenge came at the very beginning of my career. I started as Director General in 1962 with the most important goal of finding a location for the observatory. From Hamburg Observatory I was used to leading an organisation where all the framework — administration, personnel etc — was already in place. The house was already furnished when I moved in, so to speak. With ESO we had only the Convention to lean on and had to start from scratch with everything. It is hard to describe the working conditions but it was a real challenge and what happened later must be seen in that light.

In 1963 I had been asked by the ESO Council to clarify the relations with the Chilean government, and with AURA in the US, with whom we were discussing collaborating [see p. 25]. In October of that year I flew to New York for discussions with AURA. The disappointment was great as it turned out that we had big differences in our view of how the foundation of ESO should be set up. AURA wanted to work with the universities in Chile and we with the Chilean government. During a chat with Prof. Erich Heilmeier, astronomer in Santiago, in downtown New York that same day, it became clear that we could either continue the — presumably lengthy — discussions with AURA, and then later discuss with the Chilean government, or take the simpler and faster route: we could simply bypass the Americans and talk to the Chileans directly. At the end of October I went to Santiago and talked to the minister for interior relations. Greatly helped by the existing agreements they had set up with the United Nations Economic Commission for Latin America and the Caribbean (CEPAL), a full, but provisional agreement between ESO and Chile was quickly written. Advised by the German ambassador, I signed the agreement on 5 November 1963. In retrospect it was clear that without discussing this agreement with Council, I had overstepped my jurisdiction. It was naturally an extreme risk for me to unilaterally sign ESO up to setting up the observatory in Chile — Council could have fired me from my post only a year after taking up duty — but I believe it was necessary. ESO would never have taken off without this quick decision of mine. At the next Council meeting just nine days later I was reprimanded, but luckily there was no real doubt that Chile was the right home for ESO's telescopes, so soon we could focus on the next steps.

How do you see ESO's future?
Today [in 1976] ESO stands at a turning point. Over the past almost 25 years [since the first discussions in the 1950s], ESO has been going through a period of significant development. The 3.6-metre is now nearly ready and it has finally been decided that ESO should have a real headquarters building in Garching — in my opinion a decision that has come too late. ESO is now growing out of the *Observatoire de Mission* idea where an observatory just operates telescopes. It is becoming an organisation with the power to work closely with industry to develop technologies that do not exist. By bringing astronomical capacities together — people, equipment, infrastructure — a momentum can be gained which was not possible before. It will however be necessary to strengthen the training of young people, to have inspiring working conditions with lively exchange of ideas and to keep attracting young, bright minds. My vision is to follow the example of the Niels Bohr Institute in Copenhagen, which was so important for atomic physics in the 1920s and 1930s.

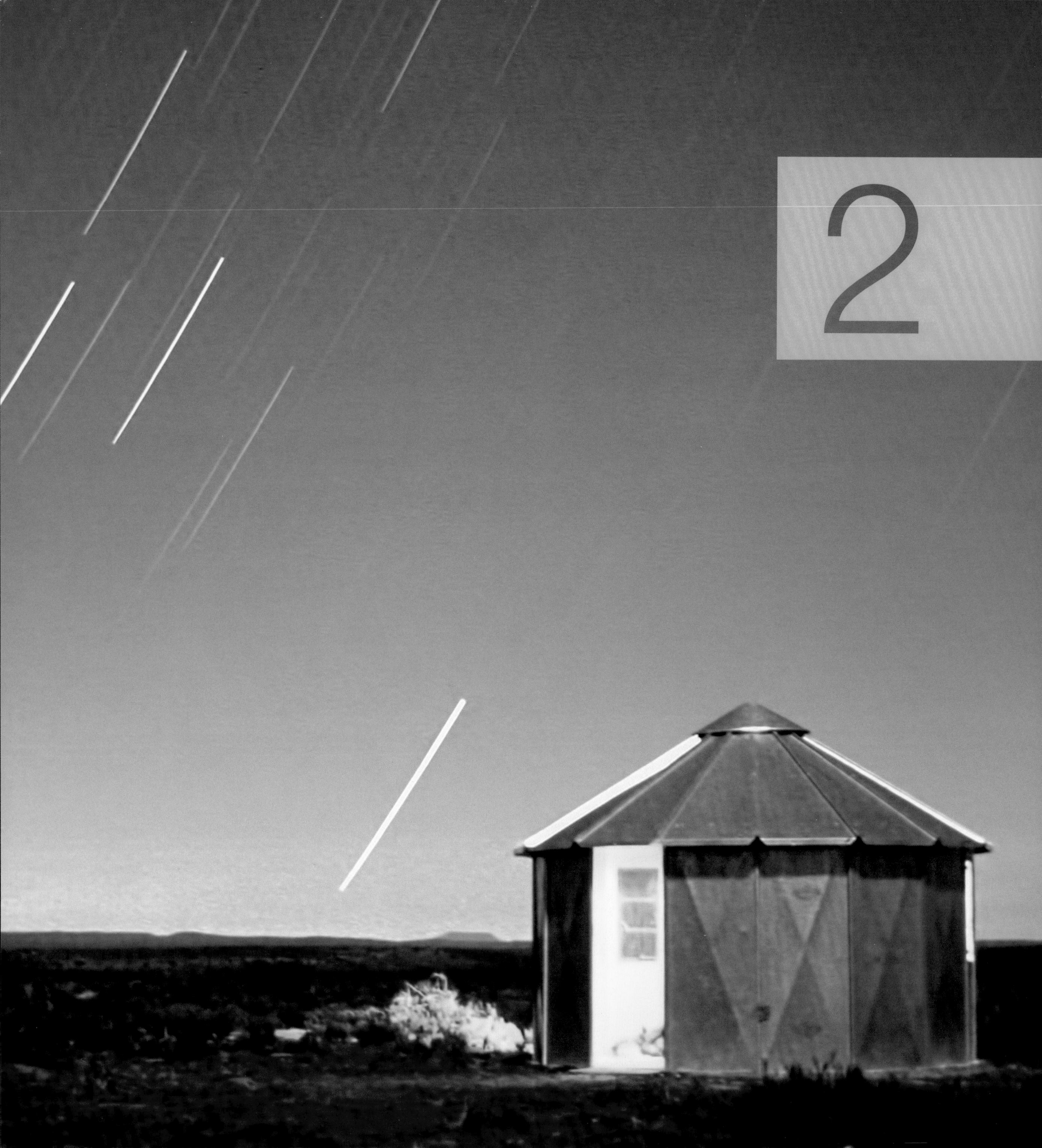

2

The Birth of ESO

European astronomers took sixteen years to turn a visionary idea into solid reality. But thanks to their commitment and perseverance, the European Southern Observatory was officially inaugurated at Cerro La Silla in northern Chile on 25 March 1969. Could Europe regain its leading role in ground-based astronomy from the United States?

In 1948, just four years prior to Jan Oort's first encounter with the southern sky, American astronomers had inaugurated the majestic Hale Telescope at Palomar Mountain in California. Its huge mirror, measuring five metres across, provided unprecedented views of planets, nebulae and galaxies. In the preceding decades, other US telescopes, notably the 2.5-metre Hooker Telescope at Mount Wilson, had already revolutionised the science of the cosmos by revealing the true nature of spiral nebulae and the expansion of the Universe. For at least half a century, America had been in the driver's seat of astronomical research.

In many ways, Europe is the cradle of astronomy. Thousands of years ago, Greek philosophers studied the skies and the motions of the planets. They charted the constellations, predicted eclipses of the Sun and the Moon, and measured the circumference of the Earth. Their fundamental premises may have been wrong — they believed that the Earth occupied the centre of the Universe, with all celestial bodies revolving around it — but this geocentric world view, written down by the great astronomer Ptolemy around 150 AD, survived with some minor additions and adaptations from Persian scientists for fourteen centuries.

Star trails over the site-testing station in South Africa
In the mid-1950s site-testing in South Africa was at its peak. An aluminium hut gives shelter during the night and is used to store the site-testing telescope during the day.

Europe is also the birthplace of the telescope

And when Ptolemy's world view was overthrown, it was another European who brought about this scientific revolution. In 1543, Polish astronomer Nicolaus Copernicus published his heliocentric model, with the Sun at the centre of the Universe. Within a few decades, Johannes Kepler from Germany, using precise measurements from the Dane Tycho Brahe, deduced the laws of planetary motion. He thus paved the way for Isaac Newton's law of universal gravitation, published in England in the second half of the 17th century. Meanwhile, it became clear that the Sun was just one of many stars in the Universe.

Europe is also the birthplace of the telescope. In 1608, when most of the United States was still unexplored, Dutch spectacle makers Hans Lipperhey and Zacharias Jansen built the very first "tubes to see far", and within eighteen months, the Italian physicist and astronomer Galileo Galilei discovered mountains on the Moon, dark spots on the Sun, the phases of Venus, the major satellites of Jupiter, and millions of faint stars in the Milky Way. Greatly improved by scientists like Christiaan Huygens in Holland (who discovered the rings of Saturn) and Isaac Newton in England (who invented the reflector), the telescope soon became the most important instrument in the study of the Universe.

Telescopes collect and concentrate starlight using convex lenses or concave mirrors. Size does matter: larger lenses or mirrors reveal fainter stars and more detail. Thus, by building larger and larger telescopes, European astronomers were able to bag one scientific breakthrough after another: the proper motion of stars, the discovery of Uranus (by William Herschel), the first estimate of stellar distances, and the spiral nature of many nebulae. Universities all over Europe established their own observatories — Leiden in the Netherlands was the first, in 1633 — and dedicated amateur astronomers like William Parsons in Ireland constructed the largest telescopes in the world.

But about a century ago, things started to change. Europe has always been a politically fragmented continent, with individual city states and kingdoms fighting for their own supremacy and prosperity; and in the field of astronomy and telescope building, no single European country could compete with the United States. There had been examples of international cooperation (the discovery of asteroids in the early years of the 19th century was the result of a pan-European search programme), but eventually, America took the lead, building bigger telescopes and attracting brilliant astronomers from the Old World.

Birth of ESO in 1953
During a boat trip in the Netherlands, Kourganoff, Oort and Spencer Jones discuss the idea of a joint European effort in astronomy.

Jan Oort and the birth of radio astronomy

Jan Oort
Oort was also fascinated by radio waves from the Universe and played a major role in starting the new field of radio astronomy.

Leiden astronomer Jan Oort not only paved the way for the birth of the European Southern Observatory; he also played an instrumental role in the birth of radio astronomy — the study of long-wavelength radio emissions from the Universe.

While the pioneering observations of cosmic radio waves were carried out in the 1930s by Karl Jansky and Grote Reber in the United States, Oort was the first to realise that radio observations might open up a whole new window on the Milky Way, partly because radio waves are not absorbed by interstellar dust clouds. In 1944, Oort's student Henk van der Hulst discovered that cold, neutral hydrogen atoms — a very important, but invisible component of the Universe — should emit at a radio wavelength of 21 centimetres. This made it possible to map the gas in the Milky Way galaxy.

The 25-metre radio telescope in Dwingeloo, the Netherlands, built on Oort's initiative in 1956, was the largest in the world for over a year. Fourteen years later, in 1970 — just a year after the inauguration of the La Silla Observatory — Queen Juliana opened the Westerbork Synthesis Radio Telescope, which is still one of the largest radio interferometers in the world. Ever since, the Netherlands has played a leading role in the field of radio astronomy, most recently with the construction of LOFAR, the Low Frequency Array.

Two World Wars didn't help either. In 1938, at the sixth General Assembly of the International Astronomical Union in Stockholm, the newly-elected President, British astrophysicist Arthur Eddington, remarked that *"in international politics the sky seems heavy with clouds, [but] such a meeting as this [...] is as when the Sun comes forth from behind the clouds. Here we have formed and renewed bonds of friendship which will resist the forces of disruption."* Within a few years, though, Europe would indeed be torn apart for the second time in the 20th century. Yet progress did resume.

In the spring of 1953, at the University of Leiden, Jan Oort discussed the future of European astronomy with the German–American astronomer Walter Baade, who had been invited by Oort to come to the Netherlands for a couple of months to prepare a conference on galactic astronomy in Groningen. That same year, European physicists were drafting the CERN convention, for close cooperation in the field of nuclear research and particle physics. Might a similar approach be fruitful in astronomy? Sixty-year-old Baade, famous for his discovery of two distinct stellar populations in the Milky Way, was enthusiastic. Before long, Oort was writing to colleagues in Belgium, France, Germany and Sweden.

On 21 June, and during the Groningen conference, the plan was discussed by leading astronomers from all over Europe, including the British Astronomer Royal, Sir Harold Spencer Jones. It sounded so obvious: a big, European observatory in the southern hemisphere — to gain access to the centre of the Milky Way and the Magellanic Clouds — equipped with a 3-metre reflector, a photographic Schmidt telescope, and a number of smaller instruments. Seven months later, on 26 January 1954, twelve astronomers from six countries met in the stately Senate Room of Leiden University to sign a statement expressing their desire to establish a European observatory in South Africa.

For European astronomers, South Africa was a logical choice. But none of the existing sites there were seriously considered as viable locations: they were all too close to major cities, and the joint European observatory not only needed good seeing, but also ultra-dark skies. In October 1955, four observers set sail to Cape Town, carrying portable 25-centimetre telescopes, and over the next couple of years, a number of new potential sites were tested, from the Johannesburg–Pretoria area in the north to the Great Karoo semi-desert in the south.

The effort paid off: it soon became clear that the southern region offered much better observing conditions. By late 1958, site-testing activities focused on the area around Zeekoegat — a small settlement north of the Groot Swartberg mountain range — and on three mountains in the Klavervlei Farm territory, some 120 kilometres further north. In subsequent years, the four isolated sites were continuously monitored by young, adventurous and dedicated people. These included Albert Bosker and Jan Doornenbal, who had been recruited from among Dutch Boy Scout leaders — quite appropriate, given the degree of sacrifice and independence that was required of the applicants.

Meanwhile, astronomers were struggling to interest their respective governments in sponsoring the endeavour. For instance, André Danjon of Paris Observatory had a hard time convincing the French administration of the necessity of the European Southern Observatory. The prospects for French participation in ESO greatly improved, however, in 1959 when the Ford Foundation in New York announced a one million dollar grant — about one fifth of the estimated capital investment for the observatory's establishment. The United Kingdom withdrew from the project in 1960, focusing instead on a Commonwealth Telescope in Australia, which later became the 3.9-metre Anglo-Australian Telescope at the Siding Spring Observatory in New South Wales.

Site-testing in the Karoo, South Africa
The weather in South Africa was good, but there were exceptions.

Was South Africa really the best possible location for the new observatory? In the late 1950s, American astronomers undertook site-testing expeditions in the rugged, mountainous landscape of northern Chile, looking for a good spot to build the southern-hemisphere counterpart of the planned Kitt Peak National Observatory in Arizona. The preliminary results were extremely promising, and in the spring of 1960, Jan Oort and acting Kitt Peak director Donald Shane seriously discussed the possibility of ESO and AURA (the Association of Universities for Research in Astronomy) sharing a Chilean mountaintop.

Eventually, in November 1962, AURA selected 2200-metre-high Cerro Tololo, some 80 kilometres east of La Serena, as the site for their future Inter-American Observatory, well before the Europeans had made up their minds. But by then, ESO had made another important leap forward of its own. On Friday 5 October 1962, at the French Ministry of Foreign Affairs in Paris, the Ministry's Secretary-General and the ambassadors of Belgium, Germany, the Netherlands and Sweden officially signed the ESO Convention. The European Southern Observatory finally became a reality. *"Alleluia"*, wrote André Danjon in a letter to his German colleague Otto Heckmann.

So what about Chile? Well, European interest had certainly been piqued. In December 1962, while site testing in South Africa was still going on, two ESO observers visited both Cerro Tololo and Cerro La Peineta, a few hundred kilometres further north. And in June 1963, a large group of ESO officials rode up to Cerro Tololo on horseback, and also visited nearby Cerro Morado. The group included ESO initiator Jan Oort, as well as Otto Heckmann, who had become ESO's first director in late 1962. On 15 November 1963, the ESO Committee unanimously decided to shift the focus from South Africa, and to build the European Southern Observatory in Chile.

But where? Working together with AURA turned out to be a dead end, because the Americans were negotiating with the University of Chile, while ESO opted for a contract at government level, and an extraterritorial status

for its observatory. In the spring of 1964, a new expedition headed by Heckmann took a close look at three new mountaintops: Guatulame (well south of Tololo), Cinchado (close to Tololo, and actually on AURA territory), and La Silla, about 100 kilometres further north. On 26 May 1964, just five weeks after Heckmann first mentioned this mountain in a letter to Jan Oort, the ESO Council picked Cerro La Silla as the site for its future observatory.

So what about an extensive site-testing campaign, as had been conducted in South Africa? Not for La Silla. The earlier US expeditions had clearly shown that virtually every peak in the area offered superior observing conditions. Moreover, La Silla was government property, which would make it much easier to acquire the land. The contract with the Chilean government was signed on 30 October 1964: an area of 627 square kilometres was purchased for just 60 000 D-Marks. A few months later, ESO bought a nice villa in Santiago's Las Condes district that would be transformed into a pleasant guesthouse for visiting personnel.

Now all that was left to do was to prepare the site for the construction of the observatory. The basic dirt track that connected the La Silla area to the Pan-American Highway was improved over the years, and from Camp Pelícano, at the base of the mountain, a new, winding road to the summit was constructed — twenty kilometres long, on average five metres wide, with no sharp curves and with a maximum slope of twelve percent. The summit road was dedicated on 24 March 1966.

Director General Adriaan Blaauw

Name: Adriaan Blaauw
Year of Birth: 1914
Nationality: Dutch
Period as Director General: 1970–1974

Note: Blaauw died on 1 December 2010. This "interview" is based on interviews he gave to American astronomy historian David DeVorkin in 1979 and with Dutch science journalist Margriet van der Heijden in 2010.

What sparked your interest in astronomy?
My interest in astronomy developed principally from reading the popular books of Camille Flammarion, the famous French populariser of astronomy. His books were translated into Dutch, very well done, with very good illustrations. I read and re-read these books. They have very strongly influenced me.

What do you recall of the birth of ESO?
It all started during a stay made by Walter Baade at the Leiden Observatory early in 1953. He discussed all kinds of things with the astronomers, including the general level and the future of European astronomy. And it was Baade who said, I think in a discussion with Jan Oort: *"If you European astronomers really want to reach a level of performance comparable to that of the United States, especially in the Californian observatories, then you ought to join forces and have one big telescope that*

you can pay for really only if you put together your financial resources and your resources of astronomers and technicians." I remember that Oort came to me, rather excited, in my Leiden office, which was just across the hall from Oort's. He said, *"Baade says we should do so and so, and wouldn't that be a good idea?"* And of course we all said that it was a good idea.

Can you describe the spirit of ESO?
I would say the spirit of ESO was already noticeable in the very early days, when a group of European leading astronomers got together and said, *"We must do this thing jointly."* Maybe one should rather say that doing things jointly is something that we did almost automatically in those early days. We felt that you had to pool all your resources in order to get something done. And maybe it is better to say that we were just little pioneers in this idea which was taken up later on a bigger scale by the politicians in the European economy. To us it just came naturally that you had to do things that way and we still feel that way.

You recently visited the Very Large Telescope in Chile. How was that?
It's almost unbelievable how this great, hyper-modern installation has been constructed here, in the most extreme and apparently uninhabitable desert. I was also struck by the enormous economic development that Chile has undergone since the 1960s. And it was a wonderful experience to meet some of my collaborators of the very early days, including the former boy scouts whom I had hired to carry out the really demanding tasks. They were the true desert pioneers, and it must have been a tremendous struggle for them. They are now retired, but still live in Chile.

What about ESO's future?
During my 96-year lifetime, the history of astronomy has completely been rewritten. Using the VLT, we can now even follow the orbital motions of stars whirling around the supermassive black hole in the Galactic Centre — isn't that fantastic? And of course, astronomers are discussing this great new project, the European Extremely Large Telescope. A few years ago I hardly could believe they were taking this idea seriously. But I'm sure it will be completed. It will take us even further into the Universe, and closer to the origin of the cosmos.

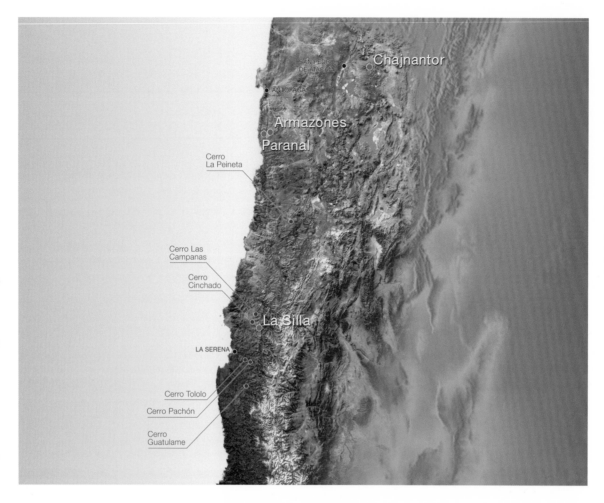

Map of the northern part of Chile
This map shows part of northern Chile with all three ESO observatory sites indicated: La Silla, Paranal-Armazones and Chajnantor. Also marked are other large observatories in Chile and some of the mountaintops investigated before settling on La Silla.

An early La Silla
This photo was taken in the late 1960s or early 1970s from the dome of the ESO 1.52-metre telescope, which had its first light in 1968. The ESO 1-metre Schmidt telescope is prominent. Behind it, at a higher level, are the water tanks of the observatory. The La Silla summit has been prepared for the 3.6-metre telescope.

A few months later, the first telescope went up to La Silla: a 1-metre photometric instrument that had been designed and constructed in the Netherlands, with a primary mirror from Jenoptik in Jena, Germany. The telescope had already arrived in Chile in the summer of 1965, where it had been sitting in a warehouse for about a year, and in 1966 it still had to be mounted in a temporary dome. But before the year was over, Jan Borgman of the Kapteyn Laboratory of the University of Groningen and his collaborators carried out the first scientific observations from the new observatory.

On Monday 25 March 1969 — at the height of NASA's Apollo programme, which would culminate with the first man on the Moon less than four months later — over three hundred guests gathered at the summit of Cerro La Silla for the official inauguration of the European Southern Observatory. The dream of Walter Baade (who had died in 1960) and Jan Oort was finally realised. Europe would reach for the southern sky, and — who knew — might gain back its leading role in ground-based astronomy from the United States.

3

In the Saddle

In the 1970s and 1980s the European Southern Observatory at La Silla became one of the largest and most productive astronomical centres in the world, with its dozen-plus telescopes scrutinising the night sky and revealing cosmic fireworks. La Silla also hosted the technological testbed for a whole new generation of large telescopes that would one day revolutionise astronomy.

Oscar Duhalde needed a break. Working as a telescope operator at the American Las Campanas Observatory in Chile, just north of La Silla, Oscar liked to go for a stroll and enjoy the spectacular view of the night sky. But this time, on the night of Monday 23 February 1987, he saw something strange — a small star in the fuzz of the Large Magellanic Cloud that didn't belong there. Meanwhile, at another Las Campanas telescope, Canadian astronomer Ian Shelton was actually taking photographs of the Large Magellanic Cloud. When he developed his plates around 02:40 the same night, he also noticed the strange star, at the edge of the Tarantula Nebula. Very soon it became clear that Duhalde and Shelton had discovered the first naked-eye supernova in almost 400 years.

The stellar explosion actually happened long, long ago, when *Homo sapiens* just started to roam the African savannah. Travelling at a speed of 300 000 kilometres per second, it took the light of the explosion some 167 000 years to cover the distance between the Large Magellanic Cloud and the Earth. When Greek philosophers first started to think about how the cosmos might be laid out, the supernova photons had already completed 98 percent of their journey. On a cosmic timescale La Silla was established just in time to observe the supernova in stunning detail.

La Silla soon after sunset
The splendours of the southern sky can truly be appreciated from the La Silla ridge. The MPG/ ESO 2.2-metre telescope is seen in the foreground.

Star trails over La Silla
A series of nighttime exposures captures these impressive star trails over ESO's La Silla Observatory. The stars appear as trails because of the apparent daily motion of the sky, which is, in fact, due to the rotation of the Earth around its own axis. The trail of an aircraft is seen over the horizon.

In 1987, La Silla (Spanish for chair or saddle, after the saddle-shaped mountaintop) was by far the most productive astronomical observatory in the southern hemisphere. As Supernova 1987A could not be observed from Europe or the United States, ESO astronomers had a ringside seat to the cosmic spectacle. So when they heard about the discovery at nearby Las Campanas, they immediately trained their telescopes on the distant explosion, which reached its peak brightness in May. For many months, on a daily basis, La Silla telescopes collected images and spectroscopic measurements, with surprising results.

For instance, Supernova 1987A was the first stellar explosion for which the progenitor star had been captured on photographic plates before it detonated. Contrary to expectations, Sanduleak −69° 202, as it was known, turned out to be a blue supergiant rather than a red giant. ESO observations also revealed the supernova to be slightly asymmetric; incandescent rings and absorbing dust shells were detected around the slowly fading glow of the explosion, and astronomers were even able to reconstruct a 3D-view of the event.

Supernova 1987A in the Large Magellanic Cloud
Image obtained with the ESO 1-metre Schmidt telescope of the Tarantula Nebula in the Large Magellanic Cloud. The bright supernova is near the middle of the image.

In 20 years La Silla had grown from a barren mountaintop with just one telescope in a temporary dome to the biggest astronomical observatory in the world, brimming with activity and sporting over a dozen telescopes, most of them heavily oversubscribed. Observing proposals from European astronomers were evaluated by an Observing Programmes Committee. Selected observers flew to Santiago, spending a night in the ESO Guesthouse before boarding a small propeller plane that took them to the Pelícano airstrip, just a short drive from the 2375-metre-high mountaintop.

It's a wonderful experience, to wind your way up to La Silla and watch the observatory glide into view — a long string of blindingly bright telescope domes dotting the gentle curve of the saddle. There are small but convenient dormitories where guest astronomers sleep during the day: *Silencio! Astronomos durmiendo!* There is time to relax with colleagues over dinner in the cafeteria, or to soak up the soothing atmosphere of the library. And when the Sun sets behind the mountain ranges in the distance and darkness slowly creeps across the silent desert, the night sky reveals its full splendour. Time to start observing.

Aerial view of La Silla
Flying along the
La Silla Observa-

In the late 1960s, a number of smallish telescopes joined the first 1-metre photometric instrument on La Silla. They included a few national telescopes, built and operated by astronomers from one of the ESO Member States, for instance the Bochum 0.61-metre telescope and the Danish 0.5-metre telescope, an almost identical copy of which was also operated by ESO. Then there was ESO's 1.52-metre spectrographic telescope, and the French 40-centimetre astrograph, officially known as the Grand Prisme Objectif telescope, which had previously carried out site-testing observations at Zeekoegat in South Africa.

Later, more and more telescopes were added: the 1-metre Schmidt telescope in 1971, the giant 3.6-metre telescope in 1976, the Danish 1.54-metre and the Dutch 0.9-metre national telescopes in 1979, the 1.4-metre Coudé Auxiliary Telescope in 1981, and the MPG/ESO 2.2-metre telescope in 1983. All these facilities had their own building, topped with classical observatory domes of white fibreglass or shiny aluminium. At night, the occasional faint glow of torchlight, the smell of coffee, and the sound of Vivaldi or Vangelis wafted up through the observatory slits, while starlight rained down, to be captured and digested by cameras and spectrographs.

ESO's 1-metre Schmidt telescope — in fact a huge camera with a field of view ten times as wide as the full Moon — played a crucial role in creating a huge photographic sky atlas of the southern celestial hemisphere. This ESO/SRC Survey, carried out in close cooperation with a similar Schmidt telescope in Australia operated by the British Science Research Council (hence the acronym), constituted the southern counterpart of the successful Palomar Observatory Sky Survey of the northern hemisphere. After being exposed for an hour or more, the huge photographic glass plates were shipped to Europe, where they were processed at ESO's Sky Atlas Laboratory, located at the premises of CERN, the European particle physics laboratory near Geneva.

Before ESO moved its headquarters from Hamburg to Garching, near Munich, Germany, in 1980, CERN was home

to the ESO Telescope Project Division, charged with realising the design and the construction of the largest telescope at La Silla. In the 1950s, Jan Oort and Walter Baade had envisioned a European 3-metre telescope, more or less comparable to the impressive Shane reflector at Lick Observatory in California. Eventually, this evolved into the current 3.6-metre telescope, in its huge 28-metre-diameter dome, located on the highest peak of La Silla. In November 1976, the giant yellow telescope on its blue horseshoe mount captured its first views of the night sky.

Like many other La Silla telescopes, the 3.6-metre telescope was constructed on the top floor of a tall building, where measurements had revealed atmospheric turbulence to be much less of a problem than at ground level. As of 1981, a smaller, separate dome housed the remotely-controlled 1.4-metre Coudé Auxiliary Telescope, which fed the collected starlight through a tunnel into the main building, where it entered the huge coudé echelle spectrometer instrument. In 1987, the instrument made headlines with its discovery of radioactive thorium-232 in stars, enabling astronomers to better constrain the ages of the oldest stars and of the Universe as a whole.

Star trails over the ESO 3.6-metre telescope
The ESO 3.6-metre telescope which today hosts HARPS, the High Accuracy Radial velocity Planet Searcher — the world's foremost exoplanet hunter.

The MPG/ESO 2.2-metre telescope
The 2.2-metre telescope has been in operation at La Silla since early 1983 and the telescope time is shared between the Max Planck Society and ESO.

Small is beautiful

In astronomy, bigger is better. At least, that's the impression garnered from the history of the telescope. Larger mirrors catch more starlight, reveal more details, and let you peer further into the depths of the Universe. But there's a downside too. Large telescopes are expensive, very much in demand, and usually observations are planned many months in advance.

When speed is of the essence, astronomers rely on small, flexible and relatively cheap telescopes, preferably fully robotic. For instance, when an energetic gamma-ray burst is discovered by a satellite observatory — a flash of high energy radiation, probably caused by the explosive death of a very massive star, or by the merger of two compact neutron stars — scientists like to carry out immediate follow-up observations with optical telescopes, to catch the afterglow of the burst before it fades back into oblivion.

Two small telescopes at La Silla offer that capability. The Rapid Eye Mount (REM) is a 60-centimetre robotic telescope in a small dome, operated by Italian astronomers. TAROT (Télescope à Action Rapide pour les Objets Transitoires) is an even smaller (25-centimetre) French telescope located in a small building with a simple sliding roof. Both instruments respond to automatic triggers from space telescopes like Swift and Fermi.

Another 60-centimetre robotic telescope is TRAPPIST (TRAnsiting Planets and PlanetesImals Small Telescope), operated by Belgian and Swiss astronomers to hunt for exoplanets that cross the face of their parent stars, and for icy bodies in the outer reaches of the Solar System.

TRAPPIST
The TRAnsiting Planets and PlanetesImals Small Telescope (TRAPPIST) is a 60-centimetre telescope at La Silla devoted to the study of planetary systems.

The Rapid Eye Mount telescope
A 60-centimetre robotic telescope in a small dome, operated by Italian astronomers. The main purpose of the REM Telescope is to follow up promptly the afterglows of gamma-ray bursts.

TAROT
The sliding roof enclosure that holds the 25-centimetre TAROT (Télescope à Action Rapide pour les Objets Transitoires). It is a very fast-moving optical robotic telescope on La Silla that can provide fast and accurate positions of any quickly evolving events on the night sky within seconds.

Nighttime at La Silla in 2011
A sprinkling of snow leaves the ground between La Silla's domes white.

**A panorama of a
unique cloudscape
over La Silla**
One of the most dra-
matic photos taken of
La Silla shows a rare
cloudscape. Located at
the southern edge of
the Atacama
Desert, this is the
home of ESO's first
observing site.

The ridge of La Silla
The ridge on top of Cerro La Silla is littered with domes and buildings. La Silla remains one of the most productive ground-based observatory sites in the world.

In 1983, the MPG/ESO 2.2-metre telescope saw first light. Built by the German Max Planck Society, the telescope is on indefinite loan to ESO. Equipped with the 67-million-pixel Wide Field Imager since 1998, this telescope has produced some of the finest photographs of the southern sky, including a breathtaking panoramic view of pink, swirling gas clouds and newborn stars in the Lagoon Nebula.

And just beyond La Silla's highest peak, invisible from the mountain road and from the main part of the observatory, Swedish astronomers erected the photogenic Swedish-ESO Submillimetre Telescope (SEST) in 1986. Like similar instruments operated by IRAM (Institut de Radioastronomie Millimétrique) at the Plateau de Bure in the French Alps, this 15-metre radio dish collected cosmic microwaves, emitted by dark clouds of cool dust and molecular gas that hardly give off any visible light.

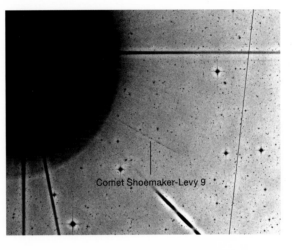

Comet Shoemaker-Levy 9
An image showing Comet Shoemaker-Levy 9 (SL9, marked) shortly before crashing into Jupiter (the big glare left). This negative image was taken with the ESO 1-metre Schmidt telescope, at La Silla in Chile.

Comet Shoemaker-Levy 9

As 1987 came around La Silla had become a mature observatory, and as the photons of Supernova 1987A completed their 167 000 light-year journey and finally arrived on Earth, ESO was more than ready to take part in one of the largest coordinated international observing campaigns in the history of astronomy. By then it had become the most productive observatory in the southern hemisphere, if not the world.

In July 1994, the whole armada of ESO telescopes was again aimed at a unique cosmic spectacle in an even larger campaign. Not an exploding star this time, but the violent crash of a whole stream of cometary fragments into the atmosphere of the giant planet Jupiter. In an unprecedented move, ESO had allocated a total of more than forty nights of observing time at half a dozen telescopes to observe the resulting explosions. And with comet Shoemaker-Levy 9 performing its dazzling exit in the early days of the internet, the data were distributed across the globe almost in real time, while ESO issued electronic news bulletins on a daily basis.

The ESO 3.6-metre telescope at La Silla
The ESO 3.6-metre telescope started operations in 1976 and set Europe the exciting engineering challenge of constructing and operating a telescope in the 3–4-metre class in the southern hemisphere. The smaller dome on the right holds the Coudé Auxiliary Telescope.

The SEST at La Silla
The 15-metre Swedish-ESO Submillimetre Telescope at ESO's La Silla observing site in the southern part of the Atacama Desert of Chile.

The Lagoon Nebula
The amazing vista of
the Lagoon Nebula was
taken with the Wide
Field Imager attached
to the MPG/ESO 2.2-
metre telescope.

The New Technology Telescope
The Sun sets behind the NTT (right) at La Silla. The Swiss 1.2-metre Leonhard Euler Telescope is seen to the left.

Meanwhile, a revolutionary new instrument had taken shape at La Silla. A 3.5-metre telescope like no other, with unprecedented properties and qualities, enabled by the use of novel technologies. Its name: the New Technology Telescope (NTT). Its aim: serving as a testbed for a new generation of astronomical facilities, much bigger and more powerful than ever. At the age of 25, ESO had grown up and was now preparing for the future, and the New Technology Telescope would lead the way.

The main mirror of ESO's original 3.6-metre telescope was about half a metre thick and weighed over ten tonnes. It *had* to be that massive — a thinner mirror would not be stiff enough and would sag under its own weight, producing distorted images of stars and galaxies. But extrapolating the traditional telescope designs to much larger instruments would lead to unwieldy structures and prohibitive costs, especially if bulky equatorial mounts were to be used, where one telescope axis is parallel to the Earth's

Director General Lodewijk Woltjer

Name: Lodewijk Woltjer
Year of Birth: 1930
Nationality: Dutch
Period as Director General: 1975–1987

What was the greatest challenge during your time as ESO's Director General?
When I was heading ESO the most important issue was to increase the number of Member States so that the realisation of the Very Large Telescope could proceed. The VLT has allowed ESO to become the undisputed world leader in ground-based astronomy.

What is your favourite ESO anecdote?
When Italy decided to become a member of ESO in 1982, a final discussion was held with Vito Scalia, Italy's minister for Scientific Research, in the Sicilian town of Taormina, to fix the precise conditions. These included the idea of slicing an existing 3.5-metre diameter glass disc into two thinner discs, with each party — ESO and Italy — getting one slice to build a thin-mirror telescope. After everything had been settled, it appeared that the minister thought that the mirror would be cut vertically, into two semi-circular pieces!

How do you see ESO's future?
The most important future projects in astronomy are expected to cost a billion euros each. Examples include the Atacama Millimeter/submillimeter Array, the European Extremely Large Telescope, and the Square Kilometer Array for radio wavelengths. With every next generation becoming substantially more costly, it is no longer enough to justify the science but also to ascertain the future willingness of society to provide the funding. In this respect warning signs abound. On a fifty-year time scale the situation in astronomy is particularly critical because the most interesting topics (e.g., the possible presence of life on Earth-like planets or gravitational waves from mergers of massive black holes or from the early Universe) will require still more expensive space-based instruments.

The planet hunters

In October 1995, Geneva astronomers Michel Mayor and Didier Queloz announced the discovery of the first exoplanet orbiting a normal star. Their discovery of 51 Pegasi b, as the planet is officially known, marked the birth of one of the most exciting research fields in the history of astronomy: the search for Earth-like planets.

Exoplanets are hard to see. They're not only far away, but also much smaller and dimmer than the stars they orbit. A planet may reveal its presence by causing tiny wobbles in the motion of its parent star. This radial velocity technique, pioneered by Mayor and Queloz, is still the most efficient way of detecting exoplanets. Moreover, it's the only method that can reveal a planet's mass.

By the spring of 1998, the Swiss team had inaugurated the 1.2-metre Leonhard Euler Telescope at La Silla, equipped with a much improved version of the spectrograph that had been used to detect 51 Pegasi b. Although the telescope is used for many other types of astronomical observations, measuring stellar wobbles induces by orbiting planets is still its main purpose.

Five years later, in February 2003, Mayor and his colleagues installed an even better spectrograph at ESO's 3.6-metre telescope. HARPS (High Accuracy Radial velocity Planet Searcher) is the most sensitive instrument of its kind in the world: it can measure stellar velocities to a precision of less than 3.5 km/h — a slow walking speed!

Using HARPS, the Geneva team has bagged an impressive number of exoplanets, including the planetary system of a star known as HD 10180, and super-Earths in the habitable zones of red dwarf stars.

The exoplanet Beta Pictoris b This artist's impression shows how exoplanet Beta Pictoris b inside the dusty disc of Beta Pictoris may look.

axis of rotation. If astronomers wanted to build substantially larger telescopes, something had to give.

The ground-breaking NTT became operational in the spring of 1989. It has a strongly curved mirror, with a short focal length, so the telescope itself is extremely compact. It is mounted on an alt-azimuth mount, with one vertical and one horizontal axis — much more compact than an equatorial mount. An alt-azimuth mount requires the telescope to rotate with continuously varying speeds around two axes at once to follow the diurnal rotation of the sky,

The NTT in its enclosure
The ESO 3.58-metre New
Technology Telescope in
its compact enclosure.

but computer control — unavailable in the past — takes care of that. And instead of a conventional dome, the NTT has a much smaller, octagonal, co-rotating enclosure.

The biggest technological breakthrough of the NTT, however, was its 3.58-metre mirror. Shaped like a meniscus — a curved surface with a uniform thickness — the mirror is only 24 centimetres thick. To prevent it from deforming under the influence of gravity, temperature changes, or wind load, it is supported by 75 computer-controlled actuators that constantly flex the mirror into the right shape. This active optics technology, pioneered by ESO optician Ray Wilson, paved the way for the construction of the much larger mirrors of the Very Large Telescope.

While the NTT, the MPG/ESO 2.2-metre telescope and the venerable 3.6-metre telescope have all been refurbished, upgraded and equipped with new sensitive instruments over the past twenty years or so, most of the smaller telescopes at La Silla have by now been decommissioned, and, in some cases, even removed from the observatory. And although a number of very exciting astronomical programmes are still being carried out at The Saddle (for an example see the box opposite), the place has become a quiet echo of the bustling activity of the late 1980s.

But with the Very Large Telescope at Cerro Paranal, hundreds of kilometres further north, playing an ever more important role, the European quest to unravel the cosmos has become more intense over the past fifteen years. So what have we learnt about the Universe so far and what is our place in time and space?

A full view of the
**La Silla mountain
from foot to summit**
At the foot of La Silla
Camp Pelicano is seen
— the base camp in
the narrow valley Que-
brada Pelicano. The
small oasis seen pro-
vides the observatory's
water. ESO installed
its original base camp
in Pelicano in the
mid-1960s. La Silla
Observatory itself is
seen at the summit.

4

Cosmic Voyage

What is this Universe that astronomers try to fathom? Where did it come from and where will it go? Here is the miraculous story of the cosmos, from beginning to end — an introduction to space and a brief history of time. It's a tale of mind-blowing proportions and intricacies, and we are an integral part of it.

The Cat's Paw Nebula seen in the infrared with VISTA
The Cat's Paw Nebula is a vast region of star formation about 5500 light-years from Earth in the constellation of Scorpius. In this magnificent image the glowing gas and dust clouds obscuring the view are penetrated by infrared light and some of the nebula's hidden young stars are revealed.

A long time ago in a galaxy far, far away...

No, hang on, that won't work. Our story starts well before there were any galaxies, let alone *Star Wars* commanders. We're going way back in time — not hundreds or thousands of years, not even millions of years, but almost fourteen *billion* years. Back to the infancy of the Universe, when matter and energy had just made their first appearance on the cosmic stage.

Welcome to the origin of space and the beginning of time. Welcome to the birth of the Universe, and to the start of a grand process of evolution of which we are an inseparable part.

The newborn Universe is a hot and crowded place. Like partygoers on a seething dance floor, subatomic particles bump into each other all the time. But unlike people, the party particles annihilate each other on collision, producing a flash of energy that subsequently turns into a whole avalanche of new particles. This is a rave of matter and energy, of creation and destruction.

It doesn't last long, though. The dance floor is stretching — the Universe is expanding. Space begets more space. Densities drop, temperatures plunge. Collisions become less frequent and less energetic, and through natural decay, particles almost entirely disappear from the scene. Within a few minutes, all that remains is a thinning ooze of simple atomic nuclei and electrons in a rapidly cooling bath of primordial radiation. Before our Universe is half a million years old — comparable to the first half-day of a human lifetime — these particles end up in neutral atoms of hydrogen and helium, the simplest and most common elements in nature.

Bok Globule Barnard 68
This view of the dark cloud Barnard 68 is quite unique as it stretches from visible light (here rendered as blue) to infrared (shown in red). It shows that dust and gas scatters blue light more, and reddens the light from the background stars.

Gone is the glow and fury of cosmic birth. The hurly-burly of the Big Bang has given way to a solemn, chilly blackness and the feeble force of gravity is gaining control. Mysterious dark matter particles clump together, despite the expansion of space. They flutter down in vast sheets, slowly flowing dark ribbons, and pile up in huge clouds. Before long, the hydrogen and helium atoms follow suit, drawn in by the dark matter's gravity, like bystanders attracted to a crowd.

Thus, the first galaxies are born — the true building blocks of the Universe's large-scale structure, supported by an invisible scaffolding of dark matter. And eventually, everywhere in space, more and more lights pop up in the darkness, like lighter flames flickering in a dark hall as a backdrop to a haunting ballad. Are these new sources the energetic radiation of gas clouds that have been heated as they fall into the gravitational abyss of newly-formed black holes? Or are they traces of the very first generation of stars? As yet, no one knows, but one way or the other, the cosmic dark ages are over. The Universe starts to shine.

Stars are easy to make. So easy, in fact, that they form independently, in unimaginable quantities — there are more stars in the observable Universe than grains of sand in all the deserts and on all the beaches of Earth. Tenuous clouds of hydrogen and helium shrink and fragment under their own gravity. The densest pockets collapse into spinning spheres of gas — stars in the making. Gravity tries to squeeze the atoms even closer together, but eventually it fails: If the collapsing cloud is big enough, rising pressure and temperature in its core cause the gas to glow, and radiation pressure puts a halt to gravitational contraction.

This is the recipe for a star: Pile up enough gas in a small enough volume, let gravity take over, and the whole thing starts to glow all by itself — hot, blue and blindingly bright if the star is massive; dull red and lukewarm if it is a lightweight. A star is born, striking a delicate balance between the inward pull of gravity and the outward push of radiation pressure. And not just one star, but trillions of them — all across nascent galaxies, and quite often forming huge globular clusters.

Colliding galaxies seen with the VLT
This striking image, taken with the FORS2 instrument on the Very Large Telescope, shows a beautiful yet peculiar pair of galaxies, NGC 4438 and NGC 4435, nicknamed The Eyes.

Spiral galaxy NGC 1232
This spectacular portrait of the large spiral galaxy NGC 1232 was the first light image from FORS1, taken in 1998.

The Carina Nebula
This spectacular pano-
ramic view shows a part
of the Carina Nebula.
The image was taken
with OmegaCAM on the
VLT Survey Telescope.

An artist's rendering of a distant quasar
This artist's impression shows how a quasar powered by a black hole with a mass billions of times larger than the Sun, may look.

Meanwhile, the galaxies are slowly clustering together in giant swarms, and occasionally smashing into each other. Just as companies grow by mutual mergers and acquisitions, galaxies grow through collisions and mergers. Small, irregular clumps of dark matter and stars meld together into majestic spiral galaxies that may contain hundreds of billions of stars, while merging spirals produce even larger elliptical galaxies.

Close encounters and galaxy mergers also stir gas clouds and shake up stars. Tidal forces pull out long streams and tails of debris; shock waves ignite violent bursts of star formation, and gluttonous black holes in galaxy cores grow ever fatter by gobbling up matter. In the process, these monstrous holes spew out jets of energetic particles and radiation, turning their host galaxies into luminous quasars that can be seen from billions of light-years away.

But this early in the history of the Universe, there's no one around to marvel at the cosmic display. The Universe began as an almost pure mix of hydrogen and helium — two elements that are way too simple to create the complex molecules needed for life. Biochemistry can't

begin until heavier elements are available too, like carbon, oxygen and nitrogen — some of the basic ingredients of organic compounds. So where's the heavy element shop?

As it happens, heavy elements come for free. All you need is a bit of patience. Little by little, in the nuclear ovens deep inside the interiors of stars, new elements are forged. Not from scratch, but from the atomic nuclear fusion that results from the tremendous pressure in a stellar core. In fact, the energy released by these fusion reactions is what keeps stars shining for billions of years. First, hydrogen atoms fuse into additional helium. Then, at a later stage of a star's life, helium is converted into carbon, and carbon atoms fuse into nitrogen and oxygen.

Massive stars take this act of cosmic alchemy even further, with the synthesis of elements like neon, magnesium, aluminium, silicon, chlorine, calcium and iron. Eventually, the star's core has transformed into a storehouse of heavy elements, covered by a thick mantle of primordial hydrogen and helium. Tucked away in the interiors of trillions of stars in the Universe are the building blocks of life.

Dramatic portrait of a stellar crib
This stunning image shows the Tarantula Nebula and its surroundings. The image was made with the MPG/ESO 2.2-metre telescope and covers one square degree on the sky (four times the apparent size of the full Moon).

Cometary tales

Comet McNaught
Astronomers at ESO's observatories in Chile were optimally placed to enjoy the show of Comet McNaught displaying a vivid coma and a lovely, sweeping tail.

Ever since the first European telescopes were erected in Chile, astronomers have used them to study comets — those tiny, frozen chunks of Solar System debris that sometimes put on a spectacular display when they pass close to the Sun and develop glowing tails of gas and dust.

While conducting an inspection of a photographic plate with the ESO 1-metre Schmidt telescope in Garching in August 1975 Richard West found a trail of a diffuse comet. This would later become the great Comet West of 1976, and the first of several great comets to be studied with ESO telescopes.

For instance, in 1996, ESO's New Technology Telescope imaged dust jets from the nucleus of comet Hyakutake; similar jets from comet Hale–Bopp were observed a year later with the MPG/ESO 2.2-metre telescope, also at La Silla.

In August 2000, the Very Large Telescope's sharp vision showed details of the break-up of the nucleus of comet C/1999 S4 (LINEAR), which had been discovered by the Lincoln Near Earth Asteroid Research programme in New Mexico.

The huge sensitivity of a giant telescope also makes it possible to observe a comet when it is much further away, and showing no activity at all. For example, in March 2003 — 17 years after it last made headlines — the extremely faint nucleus of Halley's comet was detected with the VLT at a whopping four billion kilometres from Earth. And on the eve of the launch of the European Rosetta spacecraft, in March 2004, the NTT captured the nucleus of comet Churyumov–Gerasimenko, the destination of Rosetta's ten-year journey. Observations that provided the ESA spacecraft operators with valuable clues regarding the comet's level of activity.

And don't forget: a bright comet in the velvet-black skies over the ESO sites presents an almost magical photo opportunity....

Most awe-inspiring of all of ESO's comets, however, was the great comet of 2007. Officially known as C/2006 P1 (McNaught), this extremely bright comet had a spectacular tail spanning over seventy degrees on the sky. Less impressive, but no less exciting, was the brief appearance of comet Lovejoy around Christmas 2011, after it had miraculously survived a very close encounter with the Sun.

Comet West
Comet West became a magnificent display in 1976, and the first of several great comets to be studied with ESO telescopes. It was discovered by ESO staff member Richard West.

Christmas Comet Lovejoy captured at Paranal
ESO optician Guillaume Blanchard captured this marvellous wide-angle photo of Comet Lovejoy on 22 December 2011.

And now comes the fortunate part. These chemical treasures are not locked up forever, in the innards of massive stars where they would remain as inert and inaccessible as diamonds in mountain rock. Instead, they are scattered through interstellar space when stars detonate at the ends of their lives. Powerful supernova explosions seed the Universe with the raw material of its future inhabitants. Of micro-organisms, plants, animals and people. Moreover, runaway nuclear reactions during these explosive events also create elements heavier than iron, like nickel, lead, silver and uranium.

Through nuclear fusion and supernova explosions, interstellar space becomes polluted with heavy elements. Or should we say "enriched". Even though heavy elements make up no more than one percent of cosmic matter (the rest is still hydrogen and helium), the presence of atoms like carbon and oxygen generates the full potential of astrochemistry. Molecules appear on the cosmic stage. Carbon monoxide. Methane. Water. Hydrocarbons.

Somewhere in the cosmic dark, in the outskirts of a spiral galaxy, a murky cloud of interstellar matter is disturbed by shock waves from a nearby supernova. The cloud starts to collapse under its own weight, and spawns a small cluster of new baby stars. One of them is our Sun — an inconspicuous yellow dwarf star among countless others. The Universe is about nine billion years old — almost two thirds of its present age — and finally our Solar System is born.

In nature nothing is perfect. Some two trillion trillion thousand tonnes of gas pour together to shape our Sun, but about one percent of it ends up in a flat, spinning disc surrounding the nascent star. And while this disc is mostly hydrogen and helium, it contains enough ice, dust and metals for the formation of comets, asteroids, moons and planets. Our Solar System is the byproduct of the birth of a star — cooling rubble and dirty debris on the slopes of a cosmic volcano.

In the course of a few million years, the Solar System takes shape. Chunks of ice in the cold outer reaches coagulate into the cores of future planets. With their gravity, these cores sweep up huge volumes of hydrogen and helium from the solar nebula. The end result: the four gas giants now known as Jupiter, Saturn, Uranus and Neptune. And beyond Neptune's orbit: a belt of icy leftovers, ranging in size from small comets a few kilometres across to frozen mini-worlds like Eris and Pluto.

Closer in, where ice particles evaporate because of the higher temperatures and where volatile gases are easily blown away by the fierce radiation of the newborn Sun, only a small amount of rocky and ferrous rubble survives — barely enough for the formation of a handful of smallish worlds: Mercury, Venus, the Earth and its Moon, and tiny Mars. And here, again, beyond the orbit of the outermost terrestrial planet: a belt of debris — the asteroids.

Complex organic molecules probably don't survive the energetic smash-ups of proto-planets that lead to the formation of Earth. Even most of Earth's primordial water probably gassed out and was lost to space during the planet's hot and steamy youth. But both water and organics are amply present in comets, and an early cosmic bombardment of Earth by these frozen bodies delivered a substantial fraction of Earth's oceans, and, at the same time, the carbonaceous building blocks of terrestrial life.

The Universe is a violent place. Cosmic rays wreak havoc in living cells. Supernova explosions and gamma-ray bursts sterilise nearby worlds. Long-term changes in solar output freeze or boil away planetary oceans. Killer

The Pencil Nebula
The Pencil Nebula formed from a small part of the shock wave of the larger Vela Supernova Remnant. This detailed colour composite shows thin, braided filaments that are long ripples in a sheet of glowing gas seen almost edge on. The Vela supernova remnant itself is around 100 light-years in diameter, and is an expanding debris cloud of a star that was seen to explode about 11 000 years ago.

Cosmic mysteries

The science of astronomy has made giant strides in the past century. We know how and when our Solar System formed, and we have found other planetary systems orbiting other suns. We have uncovered the births, evolutions and deaths of stars. We know about the general layout of the cosmos. We have learned how the elements in our bodies were crafted in the interiors of a previous generation of stars. We have even started to reconstruct the very early youth of our Universe. It is as if we know it all, with just a few blank spots to be filled in by future observations.

But nothing could be further from the truth. In fact, an astounding 96 percent of the contents of the Universe is a mystery — one giant question mark, staring us in the face whenever we turn our gaze to the sky. Over the past few decades, it has become clear that familiar atoms and molecules are just a minor part of the stuff that the Universe is made of. Most of the matter in space — revealing its existence through its telltale gravitational influence on its surroundings — is dark and weird, and no one knows the true nature of this dark matter.

Moreover, almost three quarters of the matter–energy content of the Universe is locked up in empty space, in the form of an uncanny vacuum energy that acts to accelerate the expansion of the Universe — a paradigm-shifting and Nobel-winning discovery made only just before the turn of the last century with important contributions from ESO telescopes. Dark energy is even weirder and more mysterious than dark matter, but many independent lines of evidence point to its existence, and it is by far the most important cosmic component in determining the ultimate fate of the Universe.

And of course there's the Big Bang — maybe the biggest mystery of all. As yet, no one knows about the very origin of the cosmos, and the scientific story of creation may well be beyond our intellectual grasp forever. So no matter how much we've learned in the past few decades, there are more than enough enigmas left for future generations of curious astronomers. There can hardly be any doubt that ESO's current and future telescopes will have a major role to play in solving a riddle or two.

The Flame Nebula
The Flame Nebula is a spectacular star-forming cloud of gas and dust in the familiar constellation of Orion. In visible light its core is hidden behind thick clouds of dust, but this VISTA image, taken at infrared wavelengths, can penetrate the murk and reveal the cluster of hot young stars hidden within.

Early days in the Solar System?
Although this artist's rendering was made to depict the scorching hot exoplanet Corot-7b, it may be close to how the Solar System appeared in its early days.

The Eagle Nebula in the infrared
Using the ISAAC instrument on VLT Unit telescope 1 at Paranal Observatory astronomers made this sharp infrared image of the Eagle Nebula, enabling them to penetrate the obscuring dust and search for light from newly born stars. The huge pillars of gas and dust in the Eagle Nebula are being sculpted and illuminated by bright and powerful high-mass stars in a nearby young stellar cluster.

asteroids produce mass extinctions. But despite these alarming threats from outer space, life on Earth miraculously survives, evolves and proliferates. Eventually, curious eyes look up at the twinkling stars overhead. Stardust returns to its roots.

As mankind longs to discover its humble place in the vastness of space and time, technology provides astronomers with telescopes to peer ever deeper into the far reaches of the Universe, and to unravel the history of the cosmos. Meanwhile, spaceflight enables humans to study other worlds up close, to leave their planetary cradle, and to set foot on the Moon. We're even dreaming of visiting the stars.

Meteorologists can't predict next month's weather. Futurologists have no clue as to events in the year 2100.

Evolutionary biologists don't know how long *Homo sapiens* will survive as a species. But astronomers do know what the distant future will bring. The Sun will slowly grow brighter and hotter, just like it has done for the first half of its life. A few billion years from now, Earth's oceans will evaporate, and the planet will turn into a greenhouse world like its hot sister Venus.

Then, as the Sun's core runs out of hydrogen, the nuclear fusion of helium will take over, and our dwarf star will turn into a bloated red giant. It will swallow Mercury, engulf Venus, and char Earth to a molten, lifeless cinder. The Sun's outer layers of gas will puff out into space, creating an expanding, colourful nebula — a tenuous shroud that will slowly dissipate into the surrounding emptiness. What's left is a puny white dwarf star, radiating its remnant heat over billions of years until it finally fades into oblivion.

"Ceci n'est pas une pipe"
Just as Magritte wrote on his famous painting "This is not a pipe", this is also not a pipe. It is rather an image of a pipe, more correctly a small part of the mouthpiece of the pipe in the Pipe Nebula. This area, also known as Barnard 59, is part of a large complex of dust and gas clouds obscuring part of the myriads of stars near the centre of the Milky Way.

The evaporated constituents of planet Earth, the elements of life, even the individual atoms of your present body, will again take part in the grand cosmic cycle of destruction and creation. Billions of years from now, they will end up in yet another cloud of gas and dust from which new stars are hatched, new planets, and — who knows — new life.

At the very end, everything comes to a halt. The ancient Milky Way will be populated with chilled-down white dwarfs, condensed neutron stars and invisible black holes — the inert corpses of former suns. Not enough interstellar matter will remain to create new generations of stars, and the accelerating expansion of space pushes galaxies away from each other, even beyond the cosmic horizon set by the finite speed of light. The Universe dies.

And then what? No one knows. Nature still harbours many unsolved mysteries. But our Universe may well be just a single act in a much grander cosmic play. Beyond the edge of space, beyond the borders of time, or, possibly, beyond the limits of our own dimensions, a multitude of other Universes may well exist, uncannily similar to or unimaginably different from our own.

Just as Earth is one of eight planets orbiting the Sun, the Sun is just one of a myriad of stars in the Milky Way, and the Milky Way is just one galaxy floating amidst billions of others in the vast cosmic ocean, our Universe could be a single facet of a brilliant, neverending multiverse. We are part of a miracle.

The Universe's biggest bangs

Gamma-ray bursts are the most powerful explosions in the Universe. Their true nature is still something of a mystery, but most astronomers believe that they are produced when the cores of supermassive, rapidly spinning stars implode into black holes at the ends of their brief lives. Jets of matter, moving at almost the speed of light, interact with stellar debris, and the resulting shock waves emit high-energy gamma rays. Producing more power in a couple of seconds than the Sun does in its entire lifespan of ten billion years, gamma-ray bursts can be observed over distances of billions of light-years by Earth-orbiting satellites like ESA's Integral and NASA's Swift.

To study gamma-ray bursts in detail, you need to be quick. The optical flash that sometimes coincides with the burst may fade away within a couple of minutes. The problem, of course, is that no one knows in advance when and where the next gamma-ray burst will occur.

That's why scientists have set up robotic telescopes that automatically respond to a spacecraft trigger. At ESO's La Silla Observatory, the 60-centimetre Italian REM telescope (Rapid Eye Mount) and the 25-centimetre French TAROT (Télescope à Action Rapide pour les Objets Transitoires) have been hunting down gamma-ray bursts since 2003. TAROT has an ultra-short response time: within one second after a gamma-ray burst trigger is received, it slews in the right direction and starts taking pictures.

Even ESO's Very Large Telescope at Paranal takes part in the action. A gamma-ray burst trigger may prompt one of the four Unit Telescopes to automatically stop its ongoing observing programme. An alarm sounds through the huge enclosure, and within ten minutes the telescope is pointed in the right direction to start scrutinising the distant explosion before it fades into oblivion.

Artist's impression of a gamma-ray burst
This artist's impression shows a gamma-ray burst in a star-forming region. Gamma-ray bursts are among the most energetic events in the Universe.

The star-forming region Messier 17
This VST image shows the spectacular star-forming Omega Nebula, also known as the Swan Nebula, as it has never been seen before. This dramatic region of gas, dust and hot young stars lies in the heart of the Milky Way in the constellation of Sagittarius (The Archer). The VST field of view is so large that the entire nebula, including its fainter outer parts, is captured — and retains its superb sharpness across the entire image.

5

The Paranal Miracle

The Very Large Telescope — ESO's astronomical workhorse for over a decade — is a smoothly running, high-tech discovery machine, perched atop Cerro Paranal in northern Chile. Sporting lasers, flexible mirrors and other optical wizardry, it is currently mankind's most powerful optical observatory — a true gateway to the Universe.

Paranal Observatory and the volcano Llullaillaco
This marvellous aerial photograph of the home of ESO's Very Large Telescope, fully demonstrates the superb quality of the observing site. In the background we can see the snow-capped, 6720-metre-high volcano Llullaillaco, located a mind-boggling 190 kilometres further east on the Argentine border.

The Sun disappears in a low, distant bank of clouds over the Pacific. Venus is already twinkling in the twilight sky, close to the thin sliver of the crescent Moon. The four shiny enclosures of the Very Large Telescope have opened, revealing the starlight-hungry telescopes inside. On the observatory platform, the long evening shadows have faded away. The small domes of the Auxiliary Telescopes still catch the pinkish glow of dusk — a stark contrast with the indigo sky on the eastern horizon, where night has already fallen on the distant, snow-covered cone of the Llullaillaco volcano.

It's a perfect, awe-inspiring mix. The cosmic drama of rotating planets and setting suns, the serene beauty of the desert landscape of northern Chile, and the impressive technology exploited by curious scientists to uncover the secrets of the Universe. Little wonder that, every evening, ESO astronomers and technicians leave the VLT control room and congregate at the platform to witness the silent spectacle. Paranal touches your heart.

At 2635 metres above sea level, Cerro Paranal, some 500 kilometres north of La Silla and much closer to the Pacific coast, is home to the workhorse of the European Southern Observatory: the Very Large Telescope. In the middle of the Atacama Desert, one of the driest regions on Earth, Paranal is an astronomical paradise, where computer-controlled optics, long light tunnels, powerful laser beams and extremely sensitive cameras and spectrographs team up to unravel the mysteries of time and space.

Sunset at Paranal
The sunset at Paranal is considered one of the magical moments you must not miss, regardless of whether you are a short-term visitor or a member of staff.

Right from the very start, visionary ESO astronomers realised that even a 3.6-metre telescope — the largest one at La Silla — would be insufficient in the long run. To determine the chemical makeup of distant Milky Way stars and to study the furthest galaxies in the Universe, you simply needed to collect much more light. And while modern electronic detectors did a much better job in terms of photon efficiency than old-fashioned photographic plates, astronomers really required bigger eyes.

In the mid-1970s, when construction of the 3.6-metre telescope was in full swing, the 5-metre Hale reflector on Palomar Mountain in California was still the largest telescope in the world, and a number of competitive 4-metre instruments had just been completed — at Kitt Peak in Arizona, at Cerro Tololo in Chile, and in New South Wales, Australia. Soviet astrophysicists were also building the 6-metre Bolshoi Teleskop Azimutalnyi, but this Caucasus colossus never lived up to expectations, and it became clear that much bigger telescopes would not be possible without changing the technology.

As early as December 1977, at a conference on Optical Telescopes of the Future in Geneva, ESO's Director General Lodewijk Woltjer, who had succeeded Adriaan Blaauw in 1975, launched his idea of building a truly gigantic telescope that would collect twenty times more starlight than the 3.6-metre telescope. Obviously, such a Very Large Telescope — the name stuck — required a number of revolutionary concepts: compact telescope design, alt-azimuth mounts, thin mirrors, active optics, co-rotating telescope enclosures and large-scale computer control.

The entrance fees of ESO's seventh and eighth Member States — Switzerland and Italy, who formally joined the organisation in 1982 — enabled the development and construction of a testbed facility at La Silla, and within a few years, experience with this 3.58-metre New Technology Telescope removed all reasonable doubt about the feasibility of Woltjer's dream. In December 1987, the ESO Council approved the construction of what would become the Paranal miracle.

The Paranal Observatory in 1999
The Very Large Telescope platform during the last stage of construction in 1999.

Early morning on Paranal
This amazing panorama shows the observing platform of the Very Large Telescope on Cerro Paranal, in Chile. Cerro Armazones is seen to the far right.

By that time, the VLT Project Group had prepared an impressive Blue Book describing the chosen design of the new monster telescope. The technological innovations offered by the NTT were leading the way, but even then, ESO's telescope designers were also following quite different routes to achieve their goal of building a 16-metre giant. Casting, grinding and polishing a delicate telescope mirror as large as a tennis court — not to mention transporting it across the globe — was of course impossible, but by the late 1970s, engineers had thought up ingenious ways to work around these hurdles.

For instance, a giant mirror could be put together from numerous smaller, hexagonal segments. This jigsaw approach was applied by the University of California and the California Institute of Technology in the construction of the first of the twin 10-metre Keck telescopes at Mauna Kea, Hawaii, which saw its first light in 1992. Or a number of smaller telescopes could be combined on the same mount — a technique that had been successfully proven in 1979 by the Multiple Mirror Telescope on Mount Hopkins, Arizona, where six 1.8-metre telescopes collected the same amount of starlight as one 4.5-metre telescope. In the case of the VLT, four 8-metre mirrors in a square pattern would mimic a virtual 16-metre giant.

But why put the four individual mirrors on the same mount? Give them each their own mount, their own enclosure, and their own set of scientific instruments, and there would be much more flexibility. The four individual 8-metre Unit Telescopes could operate separately, each observing a different target (since the full power of a 16-metre instrument isn't essential for many astronomical observing programmes), or they could join forces to collect more light from a single source. And by separating the telescopes by tens of metres, they could even team up as an interferometer — a technique we'll return to.

As for the location of the Very Large Telescope, Cerro Paranal, some 130 kilometres south of the Chilean harbour town of Antofagasta, had figured prominently on Woltjer's wish list ever since he paid his first visit to the area in March 1983. A second peak very close to Paranal was even considered as a possible site for the NTT. A few years of site testing confirmed that Paranal had even clearer skies and drier air than La Silla. Moreover, it was quite accessible, lying not too far from the old Pan-American highway. In late 1990, ESO Council gave the go-ahead and soon after, the topmost 28 metres of the conical mountain were blasted away to create a platform large enough to accommodate the new observatory.

Building four 8.2-metre telescopes posed an enormous challenge. The German glass company Schott had to construct a whole new building to cast the giant meniscus-shaped mirror blanks in specially designed rotating ovens. From Mainz, the 23-tonne mirror blanks were transported down the river Rhine, through the English Channel and up the river Seine to the REOSC polishing facility close to Paris. Each VLT mirror was then taken by sea across the Atlantic and through the Panama Canal to Antofagasta, and driven at walking pace on flatbed trucks into the Atacama Desert and up to Paranal, across tens of kilometres of unpaved roads. Meanwhile, an Italian industrial consortium was building the mechanical structures of the four VLT telescopes and their giant cylindrical enclosures.

Despite its huge diameter, each VLT mirror is just 17 centimetres thick and prone to small deformations caused by gravity, wind load and temperature changes. These are compensated for by adjustments in the computer-controlled active optics system. The supporting mirror cell contains 150 actuators; together they keep the reflecting surface to within a few nanometres of the ideal parabolic shape. But even a perfect mirror can't deliver razor-sharp

VLT's laser guide star
This spectacular image shows Yepun, the fourth 8.2-metre VLT Unit Telescope launching a powerful yellow laser beam into the sky.

Although the seeing at Paranal is among the best in the world, astronomers would love to untwinkle the stars

views of the cosmos unless it is lifted above Earth's turbulent atmosphere. And although the seeing at Paranal is among the best in the world, astronomers would love to untwinkle the stars.

And this is where *adaptive* optics comes in — a technology best described as active optics on steroids. Instead of measuring and correcting the shape of a telescope's primary mirror every minute or so, adaptive optics uses fast wavefront sensors to determine the minute distortions of starlight due to atmospheric turbulence at least a hundred times *per second*. Obviously, at that pace it's impossible to correct the shape of an 8-metre mirror precisely. But it *can* be done with a small, thin and flexible mirror in the telescope's light path, close to the focal plane. From the wavefront sensor readings, fast computers calculate the necessary corrections, and thanks to tiny actuators on its back, the small "rubber mirror" ripples in such a way as to exactly compensate for air turbulence.

To accurately measure atmospheric distortions in the direction in which the telescope is pointing, there must be a fairly bright star in the field of view. If there isn't, it has to be created. How? By shooting a thin beam of finely tuned laser light up into the air, causing sodium atoms at an altitude of some 90 kilometres up in the atmosphere to glow. Thus, an artificial laser guide star is created — too faint to be seen with the naked eye, but bright enough for the adaptive optics system.

Making waves
Here the French ESO astronomer Jean-Baptiste Le Bouquin demonstrates how waves — not light waves, but water waves in the swimming pool at Paranal's Residencia — can combine, or interfere, to create larger waves. The combination of light waves is the main principle behind the VLT Interferometer.

As if adaptive optics, with its laser guide stars, wavefront sensors and rubber mirrors isn't miraculous enough, ESO's Very Large Telescope successfully performs another trick to greatly improve its capabilities — interferometry. Known to radio astronomers for decades, interferometry is a technology that's much harder to realise at the shorter infrared and optical wavelengths. But the rewards are phenomenal: by precisely adding up the signals from individual telescopes with nanometre precision,

spatial resolution is vastly increased — astronomical jargon for image sharpness.

Interferometry is easy to explain. Suppose you had a mirror 130 metres across. Obviously, it would catch lots of starlight, and reveal ultra-fine detail. Now, paint it black, except for four circular, 8-metre wide patches, two of which are on opposite sides of the giant mirror. Even though you will have lost most of the sensitivity (as much of the big mirror is black), the spatial resolution would still be the same — and with longer exposure times, the cameras and instruments could still reveal the same level of detail. So the trick is to fool the instruments into believing that the VLT's four Unit Telescopes are part of a giant "virtual" mirror, trained at the object under study.

To achieve this, the light that is collected by two or more Unit Telescopes is fed into a network of underground tunnels. Here, one signal is delayed with respect to the other before both are combined in a high-tech interferometry lab. Because of the Earth's rotation, the required delay changes continuously, so in order to keep the signals in phase — that is, to make it look to the detectors as if both signals were reflected from mirrors at exactly the same distance from the detectors — the light beams are reflected off small mirrors mounted on carriages. The carriages move forward or backward on stainless steel tracks, their movements prescribed by accurate laser measurements.

And it's not just the four 8.2-metre Unit Telescopes that can take part in an interferometry session. To improve the efficiency of simulating a 130-metre mirror, four 1.8-metre Auxiliary Telescopes have been added to the observatory. For additional flexibility, these can be moved around to various fixed stations on the platform. Compared to their big brothers, they look cute and tiny — it's hard to imagine that before 1917, 1.8 metres was the size of the largest telescope mirror in the world.

Artificial stars

To measure the continuously changing atmospheric distortion in a particular part of the sky, a relatively bright star is needed — brighter than 13th magnitude — in the same very small field of view. Unfortunately, there are only a few million 13th-magnitude stars, so the amount of sky available naturally to adaptive optics is pretty small: typically less than one percent. The solution to this is as straightforward as it is challenging: create your own guide star.

By using a very powerful laser beam tuned to a wavelength of 589 nanometres (the characteristic emission wavelength of sodium), sodium atoms in the mesosphere are excited and begin to glow. If the 20-watt laser is focused strongly enough, this results in an artificial yellow–orange star at an altitude of some 90 kilometres, where there's a sodium-rich layer in the very tenuous upper atmosphere of the Earth. This laser guide star is too faint to be seen by the unaided eye, but bright enough to be used as a feed for the wavefront sensor.

Building a 20-watt laser is no mean feat, and tuning existing lasers to the right sodium wavelength is a challenging technique that uses frequency doublers, dyes of liquid ethanol containing organic molecules, and dye master oscillators. The laser also needs to be pulsed at microsecond frequencies, so that wavefront measurements of the laser guide star can be carried out when there is no interference from scattered laser light in the lower parts of the atmosphere.

A powerful upward pointing laser could inadvertently startle airline pilots. For safety reasons, astronomers employing adaptive optics lasers run cameras in tandem or other aircraft detection devices. They continuously survey the patch of sky surrounding the laser position. As soon as an airplane is detected, the laser is temporarily shut down.

In 2015 the VLT laser system will be expanded to use four innovative fibre lasers as part of the new Adaptive Optics Facility. A deformable secondary thin-shell mirror 1.1 metres in diameter and just 2 millimetres thick will be deformed up to a thousand times per second. This will allow much sharper images to be achieved with the HAWK-I and MUSE instruments (see box on p. 168–169) with the help of the GRAAL and GALACSI adaptive optics modules.

The galactic monster

Using the impressive light-gathering power of ESO's Very Large Telescope, in combination with the eagle-eyed vision of adaptive optics, astronomers have weighed the supermassive black hole at the core of the Milky Way galaxy: more than four million solar masses.

Since the early 1990s, a team led by Reinhard Genzel of the Max Planck Institute for Extraterrestrial Physics has used the adaptive optics instrument, NACO, on the VLT's Yepun telescope to keep track of the positions of a bunch of blue giant stars very close to the dynamic centre of the Milky Way. Over the years, these stars were seen to orbit an invisible central object, tracing out elliptical orbits with velocities of thousands of kilometres per second. In 2002, one of the stars, known as S2, approached the central object to a distance of less than 17 light-hours — only three times the distance between the Sun and the dwarf planet Pluto.

Armed with Kepler's laws of planetary motion, it is pretty straightforward to use the size and period of the orbit to calculate the central mass. The best estimate is 4.3 million times the mass of the Sun — a result that has been confirmed by Andrea Ghez's team using the 10-metre Keck telescope on Mauna Kea, Hawaii. Since the central object does not produce any observable radiation, it cannot be a dense star cluster. The only viable alternative is a supermassive black hole.

Right now, the galactic monster is pretty quiet, with just some relatively minor infrared and X-ray flickerings from its surroundings, but there is indirect evidence for bouts of much larger activity over the past centuries. The supermassive black hole in the core of the Milky Way gobbles up tenuous gas clouds, asteroid-like chunks of matter, and complete stars every now and then, producing gobbets of radiation in the process. Genzel and his colleagues have identified a huge, cold cloud of gas that might be torn apart by tidal forces when it passes close to the black hole in the summer of 2013.

Another perfect day at Paranal
Rolling red hills stretch out below the clear blue sky that is typical of Paranal. Clouds over the Pacific Ocean are seen in the distance 12 kilometres west of the observatory. Compare this photo with the one on p. 116 taken from a similar viewpoint in 1987.

Auxiliary Telescope at Paranal
The four Auxiliary Telescopes (ATs) are 1.8-metre diameter telescopes that feed light to the Very Large Telescope Interferometer at ESO's Paranal Observatory. Uniquely for telescopes of this size they can be moved from place to place around the VLT platform and are self-contained.

The Very Large Telescope is efficiently operated by well-trained observatory personnel. Gone are the days of the lone astronomer spending the night in the dome of his telescope. These days, in many cases, astronomers don't even need to travel to Chile — observations can be prepared by the scientist in advance, and carried out when the weather is optimal for that particular measurement.

But for the lucky astronomers who do actually visit Paranal, it's an unforgettable experience. From Antofagasta airport, it's a two-hour drive by rental car or ESO shuttle bus to the observatory, through a rock-strewn landscape that has an eerie resemblance to the surface of Mars. Soon after leaving the highway, the shiny VLT enclosures perched atop Cerro Paranal come into view, and it's hard to suppress the feeling that you've just entered the set of a science fiction movie.

A concrete ramp leads you down to a featureless double door under a flat dome — an uninspiring walk, as if you're about to visit some sort of underground maintenance room. But once you leave the dry and dusty desert behind, and after passing a small, dark light trap, you step into the most miraculous scene in all of the Atacama: an inviting swimming pool, surrounded by lush palm trees and tropical flowers. This is the relaxing central area of the Paranal Residencia — the partly buried hotel where staff and visiting astronomers live and sleep, work and chat, enjoy great meals and delicious Chilean sweets, go to the gym or play music. It's a true oasis in the desert; a mirage come true.

A rare sprinkling of snow
The Atacama Desert is considered one of the driest places in the world. The splendid conditions for astronomical observations in the Atacama Desert are only rarely disturbed by the weather, but when it does, it can produce unusual views of rare beauty, like here after a sprinkling of snow.

The Very Large Telescope at sunset
It is very rare to capture the VLT enclosures lit inside but this photograph was taken during routine inspection minutes after a minor earthquake. All part of the daily life at Paranal...

VISTA before sunset
During the past few
years, many astrono-
mers have arrived to
Paranal to use another
state-of-the-art tele-
scope. On a neighbour-
ing mountain, not far
from the VLT platform,
the world's largest sur-
vey telescope has been
operating since Decem-
ber 2009 — the 4.1-
metre Visible and Infra-
red Survey Telescope
for Astronomy (VISTA).

The facade of the Paranal Residencia
The rooms of the Residencia — for astronomers, engineers, and other staff working at the Paranal Observatory — face the arid Atacama Desert. In the background, the Very Large Telescope can be seen on the summit of Cerro Paranal. This award-winning construction was designed by German architects Auer+Weber.

Staff and visiting astronomers live and sleep, work and chat, enjoy great meals and delicious Chilean sweets, go to the gym or play music

Inside the Paranal Residencia
The swimming pool and the indoor garden at the Paranal Residencia. The light for this central part of the building is provided by a 35-metre dome in the ceiling. The garden and the swimming pool keep a certain level of humidity inside the building, providing a more comfortable environment to the people who work at one of the driest sites on Earth.

The Paranal base camp features technical buildings, maintenance facilities and a huge aluminising plant where the 8.2-metre VLT mirrors are recoated every 18 months or so to maintain optimal reflectivity. From the camp, it's a short but steep drive to the summit of Paranal, past the giant control room with its supporting office spaces, and up to the observatory platform. Here, you marvel at the four giant Unit Telescopes (named Antu, Kueyen, Melipal and Yepun, after the Mapuche names for Sun, Moon, Venus and the Southern Cross, respectively), at the photogenic Auxiliary Telescopes, at the VST survey telescope and the more distant VISTA infrared survey telescope, and, above all, at the terrific views of the surrounding landscape, from the quiet of the Pacific Ocean to the distant Andean peaks at the Argentine border.

And before night engulfs the observatory, before starlight rains down on the giant mirrors, and before the Very Large Telescope comes alive with all of its optical wizardry, there's ample time to witness the spectacle of the setting sun and to contemplate the miracle that European astronomers have realised here at Paranal.

Director General Harry van der Laan

Name: Harry van der Laan
Year of Birth: 1936
Nationality: Dutch
Period as Director General: 1988–1992

What makes ESO special?

ESO is a service organisation of and for its user community. That astronomical community is now uniquely collaborative. I wanted to enhance the interactive dynamics and achieved this among other measures with the introduction of Key Programmes, of the Student Programme and especially by fostering the community role in designing, building and commissioning instrumentation for the Very Large Telescope.

What was the greatest challenge during your time as ESO's Director General?

The big challenge was to make the engineering design and work out the European tendering for all subsystems of the VLT, while simultaneously running La Silla as the world's major optical/infrared observatory. The most painful restriction was not the financial budget but the staff number ceiling imposed by the ESO Council. It made life very tough for all staff and was only resolved in my successor's time.

What is your favourite ESO anecdote?

At the Council meeting of 8 December 1987, a few weeks before I took office, a long and arduous process for VLT project approval was to be completed. In a very tense atmosphere, after complex discussions, the President Kurt Hunger, started the last *tour de table* for confirmation of each delegation's commitment and green light. The last delegate to speak was Signore Griccioli, the Italian senior civil servant who normally was a jolly participant in Council debates. Now, sitting a couple of chairs to my left, he looked very gloomy, and instead of nodding approvingly like everyone else had done, he cleared his throat to ask *"Signore Presidente, what about...?"* Christian Patermann, the German governmental delegate who had worked so hard for success, sitting across from me at the wide table, jumped up, a bundle of nerves, but Griccioli looked unfazed as he continued *"... what about the press release?"* Patermann collapsed with relief, joined by roaring laughter from the whole assembly as stress released and joy broke through. The VLT was go.

How do you see ESO's future?

ESO's future depends on its community and on the future of humanity. With the Swedish-ESO Submillimetre Telescope on La Silla, ESO broke out of the optical/infrared wavelength restriction, and its key role in ALMA has confirmed that step. Now the European Extremely Large Telescope must be built. With its organisational strength and its relative financial continuity, ESO seems the best European organisation to play an ALMA-like role in the realisation of the Square Kilometer Array radio observatory, further extending its wavelength range and maintaining its unique thrust for European astronomy. Of course, none of this will happen if humanity continues on its reckless fossil fuel path, thickening the globe's CO_2 blanket to render Planet Earth less and less habitable for nine billion humans.

**Almost like being
on Mars**
Located at 2600 metres
altitude, the Paranal
Observatory sits in
one of the driest and
most desolate areas
on Earth, in Chile's
Atacama Desert. The

6

The Soul of ALMA

At 5000 metres above sea level in the Chilean Andes, the ALMA observatory is taking shape. An international partnership of Europe, North America and East Asia, the giant antenna array will help astronomers unravel the origin of galaxies, stars, planets and life. With ALMA, ESO has embarked on the biggest adventure in ground-based astronomy.

A four-wheel drive pickup truck climbs the long, steep and winding road that leads to the Llano de Chajnantor, passing close to giant cacti and watched by curious vicuñas. The view across the blindingly white Salar de Atacama, the distant Domeyko mountain range and the Licancabur and Láscar volcanoes is unforgettable. The Chajnantor plain itself, at an altitude of 5000 metres, is surrounded by other volcanoes, including dark Cerro Negro and sulphur-rich Cerro Toco. The yellows and oranges of the almost surreal landscape contrast strongly with the dark indigo hue of the crystal-clear sky. And glistening in the harsh sunlight are the dozens of dishes that comprise the ALMA observatory, located on the roof of the world, almost at the edge of space.

ALMA (Spanish for "soul") is the Atacama Large Millimeter/submillimeter Array. Expected to be completed in 2013, ESO's newest observatory is an exercise in superlatives. ALMA is by far the largest, most expensive and highest astronomical array ever built on the ground. It consists of 66 individually transportable antennas, most of them twelve metres in diameter. Together, they provide astronomers with a unique window on the coldest places in the Universe, where other suns and other Earths are born and where the building blocks of life are cooked up by organic chemistry in interstellar space.

The southern Milky Way above ALMA
The antennas of the Atacama Large Millimeter/submillimeter Array, set against the splendour of the Milky Way.

The 1950s saw the birth of radio astronomy, and in the early 1980s, infrared astronomy came of age. But the intermediate part of the electromagnetic spectrum, with wavelengths on the order of a millimetre, remained *terra incognita* for a long time, mainly because of the lack of suitable receivers. Still, astronomers realised the importance of millimetre and submillimetre astronomy for a better understanding of the origin of galaxies, stars, planets and life. Given the fact that most of today's large telescopes are well suited to observe in the near infrared, it shouldn't come as a surprise that the European Southern Observatory is now also setting its sights on the submillimetre sky.

ALMA is certainly not the first submillimetre telescope. In 1987, British, Dutch and Canadian astronomers for instance teamed up to build the 15-metre James Clerk Maxwell Telescope on Mauna Kea, Hawaii. Around the same time, the Institut de Radio-Astronomie Millimétrique established a six-antenna observatory at the Plateau de Bure in the French Alps. The United States has its own Combined Array for Research in Millimeter-wave Astronomy in California, and in the southern hemisphere, the 15-metre Swedish-ESO Submillimetre Telescope was in operation at the La Silla Observatory between 1987 and 2003. And on Pampa La Bola, near the Chajnantor Plateau, the two Japanese-led submillimetre telescopes, ASTE and NANTEN have been built. But all of these facilities are vastly overshadowed by ALMA.

Microwaves — radio waves with wavelengths as short as one millimetre — are easily absorbed by water molecules. That's why water-rich food heats up so quickly in a microwave oven. It also means that millimetre waves from deep space are absorbed by water vapour in the Earth's atmosphere. In fact, hardly any cosmic microwaves make it to sea level. To observe millimetre and submillimetre radiation, you need to be as high and dry as possible, well above much of the atmosphere, and preferably with no drop of water between the antenna and outer space. At an altitude of 5000 metres, ALMA beats all the other submillimetre telescopes in this respect and it is much larger and much more flexible to boot.

In the 1990s, various institutes developed ideas for the construction of a large millimetre-wave observatory, consisting of multiple antennas. The United States was planning a huge MilliMeter Array (MMA); ESO had its own Large Southern Array (LSA) on the drawing board, and Japanese astronomers had dreamt up an ambitious Large Millimeter Array. Over the years, the three projects merged to become the international ALMA observatory, with additional participation by Canada, Taiwan and South Korea and Chile as host state. An official agreement between ESO and the US National Radio Astronomy Observatory (NRAO) was signed in February 2003; the National Astronomical Observatory of Japan (NAOJ) joined in September 2004.

By that time, and after a long site-testing campaign, everybody agreed that the Llano de Chajnantor would be the preferred location for the new array. Lying close to the intersection of Chile, Bolivia and Argentina, the elevated plateau is large and relatively flat, and except for some rain and snow during the "Altiplanic winter" in January and February, it enjoys about nine months of ultra-clear weather under an ultra-dry sky. Obviously, constructing the necessary infrastructure would be a tremendous task, and operating a high-tech facility at five kilometres above sea level would pose many unexpected problems, but by the autumn of 2009, the first three ALMA antennas were in place.

To gain the necessary experience with high-altitude millimetre-wave astronomy, the Atacama Pathfinder EXperiment was erected at Chajnantor in 2005. APEX is a parabolic 12-metre prototype antenna for ALMA, consisting of 264 panels with a surface accuracy of just 17 micrometres — a fifth of the diameter of a human hair. It's a joint project between the Max Planck Institute for Radio Astronomy in Germany, the European Southern Observatory, and the Onsala Space Observatory in Sweden. Its detectors and receivers, including a submillimetre camera cooled to just 0.3 degrees above absolute zero, are sensitive to cosmic radiation with wavelengths between 0.2 and 1.5 millimetres, and since its inauguration, APEX has produced a steady stream of exciting results.

Star trails over APEX Although this image might at first look like abstract modern art, it is the result of a long camera exposure of the night sky over Chile's Atacama region. As the Earth rotates toward another day, the Milky Way stars above the Atacama Desert blur into colourful streaks. The APEX telescope in the foreground, meanwhile, takes on a dreamlike quality.

The promise of ALMA

ALMA will detect the glow of warm dust in galaxies further away, and thus earlier in time, than any we can detect in the deepest visible- and infrared-light photography. Further information about the early Universe may come though spectroscopic observations of carbon isotopes, since the mix of isotopes produced in stars over cosmic history is expected to evolve.

ALMA will look deep into star-forming clouds, detect the faint light emitted by infalling matter that is just starting to heat up, and map the motion of that matter.

ALMA will study all phases of planet formation. It will probe protoplanetary discs in high resolution. It may be able to detect the light from growing and warming protoplanetary cores, and to directly detect giant planets clearing paths through the surrounding discs. ALMA will be able to find even more planets by measuring the exquisitely small effects they have on the motion of the stars they orbit, and to examine the dusty debris discs that remain around stars once the gas has been removed.

ALMA will have an unprecedented ability to discover and measure the presence of molecules and their distribution in interesting structures in space. We will learn about the chemistry of space, a chemistry that can't be reproduced in laboratories on Earth, and the evolving conditions that drive it.

ALMA will investigate the great eruptions (flares) that occur on the Sun and the high-speed particles that are emitted. It will study the structure and evolution of solar prominences and filaments, strands of 6000-degree gas suspended in the Sun's three million degree atmosphere (corona).

ALMA will image planets and measure their winds. It will analyse the molecules emitted by comets and asteroids even when they're at their most interesting and active, passing near the Sun — a time when other telescopes must turn their gaze away.

ALMA will discover thousands of new Kuiper Belt objects (the class of worlds to which we now know Pluto belongs), observing the light that they emit, not their reflected sunlight. This will let us calculate their true sizes.

Last but not least, ALMA will enable us to see aspects of the Universe whose existence we didn't even suspect. Whenever we advance our abilities to capture and analyse the ceaseless stream of incoming photons from the sky, the Universe reveals new secrets. ALMA's greatest discoveries will be the ones we cannot foresee.

Four of the first ALMA antennas
Four antennas of the Atacama Large Millimeter/submillimeter Array gaze up at the star-filled night sky, in anticipation of the work that lies ahead. The Moon lights the scene on the right, while the band of the Milky Way stretches across the upper left.

First European ALMA antenna on its way to Chajnantor
The first European antenna for ALMA being transported to the observatory's Array Operations Site.

Using APEX, Per Bergman from Onsala Space Observatory in Sweden and his team made the first discovery of hydrogen peroxide in interstellar space — molecules consisting of two hydrogen and two oxygen atoms. They probably form on the surfaces of dust grains, and may facilitate the formation of water molecules. APEX has also shown us an expanding bubble of hot gas, blown into space by a superluminous star in its centre, causing colder, surrounding material to fragment and collapse into dense clumps that will hatch new stars in the near future. And in a remote galaxy, the light of which has taken some ten billion years to reach us, Mark Swinbank from Durham University in the UK found stellar nurseries that produce new stars at a prodigious rate, even though they are about the same size as the more sedate star-forming regions in the Milky Way galaxy.

ALMA will study similar things, but it is much larger and more versatile. Once completed, the array will consist of 25 European and 25 North American antennas, each 12 metres in diameter. Together, they act as a giant interferometer, not unlike the Very Large Array radio observatory in New Mexico, but sensitive to a much shorter wavelength, and so presenting a wholly different set of technical challenges. In addition, sixteen Japanese antennas (four 12-metre dishes and twelve 7-metre ones) constitute the Atacama Compact Array, which can be used separately, or as an add-on to the main array. With a combined antenna surface as large as a football field, ALMA is by far the most sensitive millimetre-wave observatory ever.

But surface area is just one aspect. ALMA's impressive flexibility is due to the fact that its individual antennas can be moved around, to create a variety of possible array configurations. When close together, within an area a few hundred metres across, the array has a relatively low spatial resolution. In other words: it can observe extended objects, but not in extreme detail. When far apart, spread out over an area 16 kilometres in diameter, the ALMA antennas provide a much sharper view.

Interferometry only works if the distances between individual antennas are known very precisely — to within a fraction of a millimetre. To achieve that accuracy, 192 antenna pads are distributed across the extended Chajnantor Plateau, like high-precision docking stations for the antennas, all connected to each other and to a powerful central computer by fibre optics. Depending on their observing proposals, astronomers can choose between 28 different array configurations. Some of the Japanese antennas of the Compact Array can even be moved around a little bit.

The cool clouds of Carina
Observations made with the APEX telescope in submillimetre-wavelength light (orange) reveal the cold dusty clouds from which stars form in the Carina Nebula.

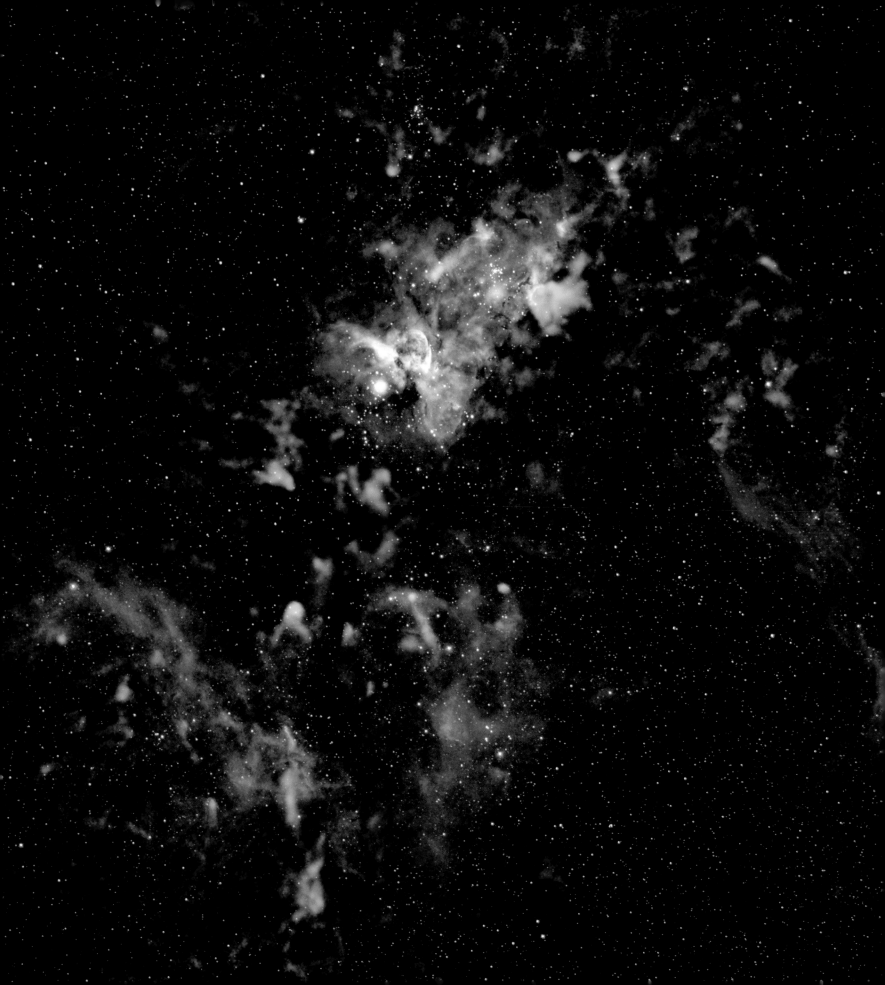

ALMA at night
This panoramic view of the Chajnantor Plateau shows the antennas of ALMA ranged across the unearthly land-scape. These crystal-clear night skies explain why Chile is the home of not only ALMA, but also several other astro-nomical observatories.

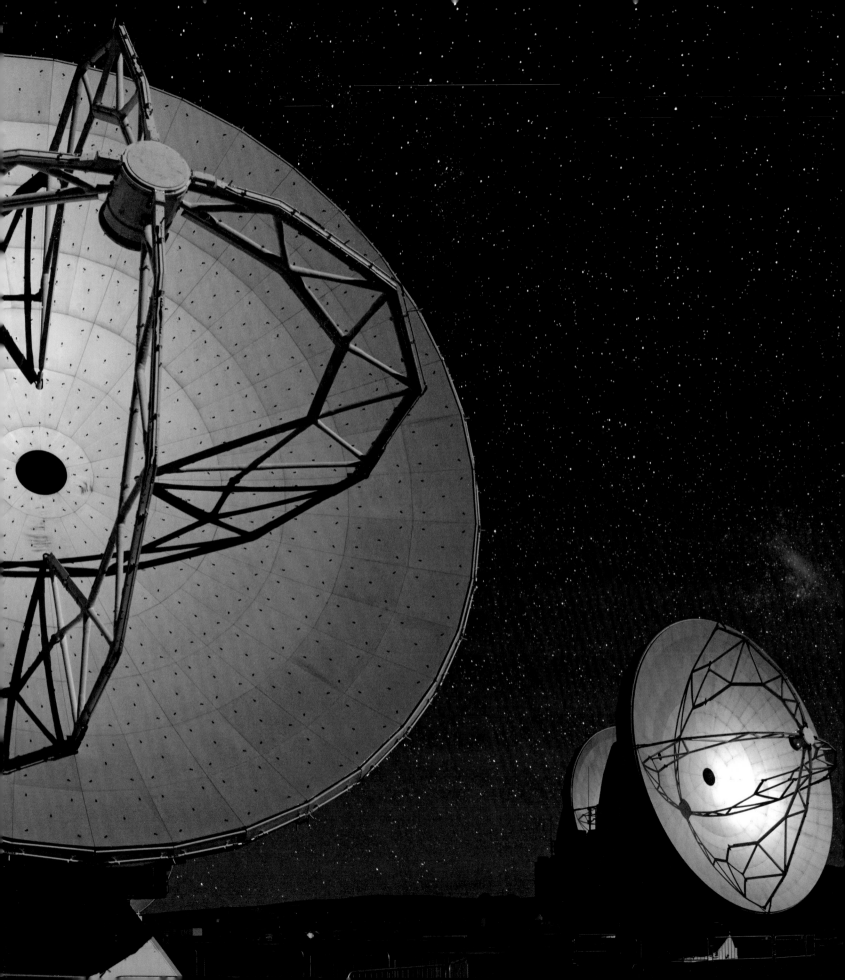

ALMA's solitude
ESO Photo Ambassador Babak Tafreshi has succeeded in capturing the feeling of solitude experienced at ALMA. ALMA's antennas appear strangely small among the peaks of the Andes.

The Moon and the arc of the Milky Way Numerous giant antennas dominate the centre of the image. When ALMA is complete, it will have a total of 54 of these 12-metre-diameter dishes, as well as 12 smaller ones. Above the array, the arc of the Milky Way serves as a resplendent backdrop. When the panorama was taken, the Moon was close to the centre of the Milky Way in the sky.

ESO architecture

ESO's Headquarters
The ESO Headquarters in Garching bei München, Germany.

It is a little known fact that ESO is also home to architectural pearls. The Headquarters building in Garching bei München, conceived by architects Hermann Fehling and Daniel Gogel from Berlin, is the scientific, technical and administrative centre for ESO's operations, and the base from which many astronomers conduct their research. The modernist building was inaugurated in 1981, and is based on a very special concept that has received worldwide attention in architectural circles. On the occasion of the inauguration an article in *The ESO Messenger* read: *"A short period of familiarisation was needed during which everybody lost their way during a few hours or a few days in what, at first sight, looks like a labyrinth."* Meanwhile ESO staff have found their way through the unusual prize-winning building, and a new spectacular expansion in glass, steel and concrete is underway.

The living quarters at ESO's Paranal site are another example of unusual ESO architecture (see the photo on p. 106–107). To make it possible for scientists and engineers to live and work at Paranal, a hotel or Residencia was built in the base camp, allowing them to escape from the arid environment outside. Returning from long shifts at the VLT and other installations on the mountain, here they can breathe moist air and relax, sheltered from the harsh desert conditions. The award-winning construction in concrete, steel, glass and wood was designed by German architects Auer+Weber+Assoziierte as a subterranean L-shape, with a 35-metre dome covering an indoor garden. The use of natural materials and colours integrates the building smoothly into the Atacama landscape. The breathtaking building has seen its share of visiting kings, princes, princesses, presidents and even James Bond himself (see. p. 124-125). The Residencia has appeared in many books on architecture such as *Pulso: New Architecture in Chile* where Kenneth Frampton poetically wrote: *"Some 200 metres below this summit a single four storey slab of a hotel, rendered red, extends itself into the limitless red desert, like the pristine relic of an ancient, alien civilization on the surface of the Moon."*

Other noteworthy pieces of architecture are the ALMA Santiago Central Office building, opened at ESO's Vitacura premises in 2010 and the ALMA hotel designed by Finnish architects Kouvo & Partanen with a planned completion date of 2014.

The souls of ALMA
The people working for ALMA are in many ways the real souls of the project. Here employees from all three partners, Europe, North America and East Asia, as well as the host nation Chile are seen.

Two giant transporters have been designed and constructed by the German Scheuerle Fahrzeugfabrik to move the delicate and expensive 100-tonne antennas around. Called Otto and Lore, each 130-tonne hauler is ten metres wide, twenty metres long and six metres high. They have 28 wheels and are powered by two 500-kilowatt diesel motors. Some 80 kilometres of roads have been constructed at Chajnantor, connecting the 192 pads. And of course, Otto and Lore also take care of transporting the antennas from the Operations Support Facility (OSF) at 2900 metres to the Array Operations Site at 5000 metres — a 28-kilometre trip.

Around the turn of the century, the only way to reach the Llano de Chajnantor was by driving cross-country from the Camino de Paso de Jama, the highway from San Pedro de Atacama to the Argentina border. But ALMA now has its own 43-kilometre access road, starting at Highway 23 between San Pedro and Toconao, with the OSF located at kilometre post 15. Housing some 500 workers, engineers and scientists, the OSF is nothing less than a small village, with its own water supply, power plant and regular bus

services to San Pedro and to the small airport of Calama, some 100 kilometres to the northwest.

At the OSF, the giant ALMA antennas are assembled and tested, in separate areas for the North American, European and Japanese contractors. Once accepted by the Joint ALMA Observatory — the international organisation that coordinates and operates the facility — the completed antennas are transported to the high site, to be added to the expanding array.

But not all support work can be carried out at the OSF. At 5000 metres above sea level, close to the heart of the array, the ALMA Technical Building has been constructed, with a giant control room overlooking Chajnantor. This North American contribution to ALMA is the second-highest-altitude steel frame building in the world. It's also home to the ALMA correlator — the supercomputer that combines the signals from the individual antennas into one coherent measurement. And for the convenience of the ALMA operators and technicians, the building is pressurised at 750 millibars — the air pressure at the OSF.

The volcano over ALMA
This impressive panoramic image depicts the Chajnantor Plateau with the majestic Licancabur volcano in the background — the ruler of the Chajnantor Plateau. *Penitentes* appear like an ice forest, disturbing the solitude of the Atacama Desert.

Outside, the atmospheric pressure is only 550 millibars, and even local Chilean workers, who are used to high altitude, are urged to use bottled oxygen to prevent physical distress.

Construction of ALMA started in 2003, and the first (North American) antenna was completed in 2008. By late 2009, the first three antennas had been transported to the AOS; the first European antenna followed in July 2011. And in early October 2011, with some twenty antennas in place, ALMA officially opened its eyes, with the start of the Early Science operations. At that time, even with a limited number of antennas in a limited number of array configurations, ALMA was already the best telescope of its kind. Over 900 proposals from astronomers all over the world had been submitted for this first phase, from which only some ten percent could be accommodated. To mark the start of the Early Science phase, an impressive submillimetre image of the Antennae galaxies was released, obtained as part of the preceding test programme.

Eventually, ALMA will play a major role in solving some of the outstanding puzzles surrounding the origin and evolution of galaxies, stars and planets. It will be sensitive enough to detect radiation from the first generation of galaxies, born a few hundred million years after the Big Bang, in the infancy of the Universe. This radiation was originally emitted at visible or near-infrared wavelengths by the primeval galaxies, but has been stretched to the longer wavelength millimetre and submillimetre waves by the expansion of the Universe during its several billion-year-long journey to Earth.

Closer to home, ALMA will shed light on the birth of stars like the Sun. The final gravitational collapse of a cool cloud of gas and dust into a dense clump that will subsequently turn into a star has never been observed before. ALMA's view will also be sharp enough to detect the circumstellar discs of gas and dust that are expected to surround every newborn star — the breeding grounds of planets. For nearby stars, it will even be possible to detect planets-in-the-making within these protoplanetary discs. And at the end of a star's life, ALMA should be able to probe the dusty stellar winds that are blown into space — the building material for a future generation of planets in other solar systems.

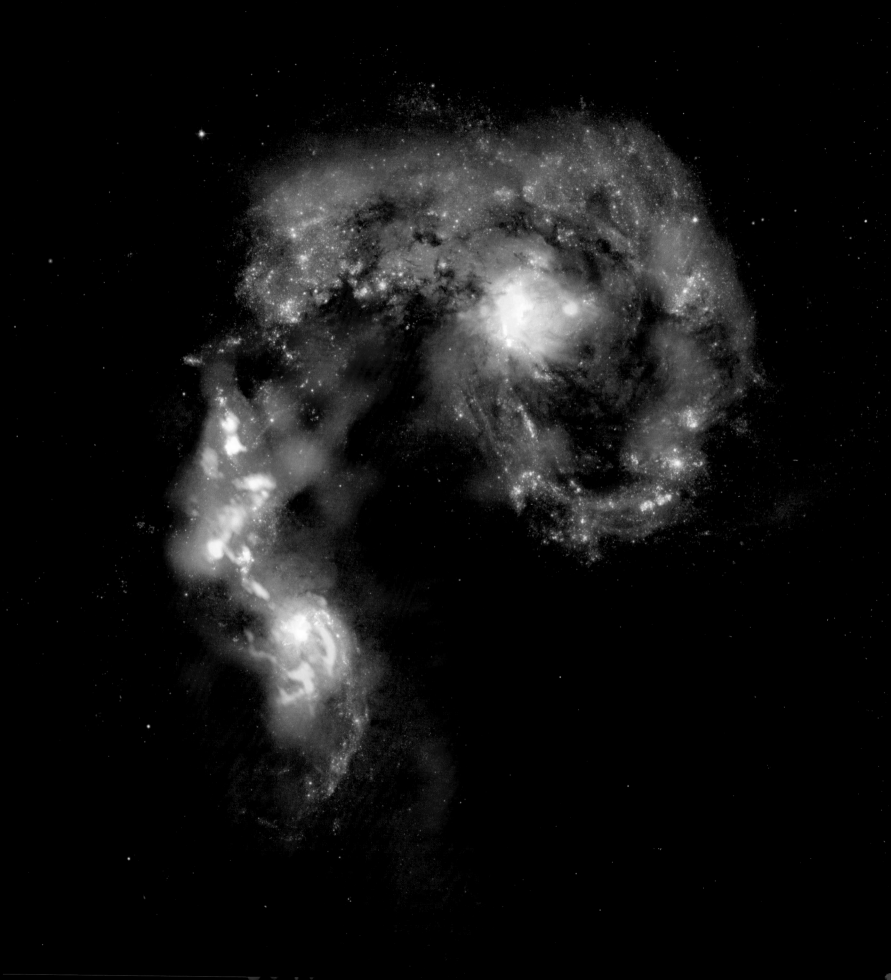

Flares and filaments on the Sun, planetary atmospheres, the sulphur volcanoes of Jupiter's moon Io — the Solar System will also be studied in detail by the giant array observatory. But the biggest breakthroughs may be in the field of astrochemistry. So far, astronomers have detected almost 200 species of interstellar molecules, and ALMA is the best possible instrument to study their properties and behaviour in much more detail, and to discover new ones. Learning about cosmic chemistry is a prerequisite to understanding the origin of organic molecules, hydrocarbons, amino acids and life.

Once in full operation, ALMA might change the landscape of astronomical research forever, and the European Southern Observatory is part of the adventure. In fifty years, ESO has grown from an idea discussed on a boat trip in the Netherlands to an ambitious, pre-eminent science and technology organisation. Together with its partners it operates the largest ground-based astronomical facility in the world, in a wavelength region that astronomers hardly knew anything about in 1962.

Chajnantor

The Llano de Chajnantor (the Chajnantor Plateau), home to the Atacama Large Millimeter/submillimeter Array, is a relatively flat area at an average elevation of some 5000 metres, close to the intersection of Chile, Argentina and Bolivia. It is part of the geologically very young Purico Complex — a 20 by 30 kilometre pyroclastic field.

The Llano de Chajnantor is surrounded by volcanic cones, including the stratovolcanoes Cerro Purico (5703 metres), Cerro Macón (5130 metres) and Cerro Toco (5604 metres) and the lava domes Cerro Agua Amarga (5058 metres), Cerro Áspero (5262 metres), Cerro Chajnantor (also known as Cerro Cerrillo, 5639 metres), Cerro El Chascón (5703 metres), Cerro Negro (5016 metres) and Cerro Putas (5462 metres). A lower part of the plateau, a few kilometres northeast of Llano de Chajnantor, is known as Pampa la Bola.

Because of the high altitude and the corresponding low levels of atmospheric water vapour, Chajnantor is one of the best places in the world for far-infrared, submillimetre and millimetre astronomy. While ALMA is being constructed on the main plateau, at some 5000 metres above sea level, other observatories are operated or planned at even higher altitudes. At 5190 metres, on the west flank of Cerro Toco, sits the international Atacama Cosmology Telescope (ACT, led by Princeton University), a 6-metre instrument studying the cosmic microwave background. Close to the ACT is the Huan Tran Telescope, a similar but smaller (3.5-metre) project of the University of California at Berkeley. At the summit of Cerro Chajnantor is the 1-metre mini-TAO telescope (Tokyo Atacama Observatory) — a pathfinder instrument for a future 6.5-metre optical-infrared telescope of the University of Tokyo. Close to the TAO site, a group of international universities are planning a giant 25-metre submillimetre telescope known as the Cerro Chajnantor Atacama Telescope (CCAT, formerly known as the Cornell-Caltech Atacama Telescope).

Icy *penitentes* by moonlight on Chajnantor
These bizarre ice and snow formations on Chajnantor are known as *penitentes*. They are illuminated by the light of the Moon, which is visible on the right on the photograph.

The completed ALMA array on Chajnantor An artist's rendering showing the final ALMA array.

Director General Catherine Cesarsky

Name: Catherine Cesarsky
Year of Birth: 1943
Nationality: French
Period as Director General: 1999–2007

What makes ESO special?

What makes ESO special among scientific organisations is that it deals with a very special discipline, astronomy, which nurtures human culture and has enormous appeal to the public. ESO shares with other European intergovernmental organisations the advantage of constructing and operating what is recognised, at the world level, as prime facilities in its field.

What was the greatest challenge during your time as ESO's Director General?

There were quite a few big challenges, but perhaps the greatest one was, after the OWL review (Over-Whelmingly Large telescope) in November 2005, to arrive at a basic reference design for the European Extremely Large Telescope that I found satisfactory from the point of view of performance, technical innovation and cost, and that would be fully supported by the community. The recipe was simple: maximum involvement of the community on the one hand, and reliance on the outstanding expertise of the ESO staff and of a number of European laboratories on the other. I remember writing to about fifty scientists and engineers, on 22 December 2005, asking them to participate in this endeavour; by the end of the year, almost all had accepted. By April 2006, all the working groups I had set up had delivered their first report, and by the end of November 2006 we could hold a large meeting in Marseille where the design was established.

What is your favourite ESO anecdote?

I have fond memories of the visit made by Chilean President Ricardo Lagos to the Paranal Observatory in 2006, with his wife Luisa Durán and with a Chilean astronomer (a great friend of his and mine), Maria Teresa Ruiz. I flew from Garching, Germany, to Santiago for this event, and arrived there, on the morning of the visit, in my usual travel gear: old blue jeans and a sweat shirt. After a while, it became obvious that my suitcase had not arrived. Together with Mary Bauerle from the ESO Santiago office, we rushed to a horrible shopping mall not too far from the airport, and swiftly bought low quality, but presentable clothes, shoes and toiletries, Mary getting half of the stuff while I got the rest. We made it just in time for me, in suitable gear, to take the plane to Antofagasta with Luisa Durán. In Antofagasta, we took a military plane to Paranal, and this was a rare opportunity to fly over our beautiful observatory. There were strict rules about remaining seated in this military plane, but when the plane made a special tour over the Very Large Telescope, Ricardo Lagos and I released ourselves and stood up to best enjoy the view, chided by Luisa Durán. The presidential couple spent the night at the Paranal Residencia; the visit was highly enjoyable, in a most friendly atmosphere.

How do you see ESO's future?

First of all, I believe in the future of astronomy; there is still so much to discover and understand! So, I hope and expect that ESO will continue to fulfill its mission of providing astronomers with first-class facilities for a long time. As an organisation, it will become more open to the rest of the world, and also may continue to further broaden the wavelength range covered. The relationship with space astronomy and space instrumentation may also be enhanced in the future. Fifty years from now, it should be a different ESO, run in ways we cannot yet imagine, riding — as it is now — on the continuous progress of science, technology and communications.

Southern star trails over ALMA
This picture may easily evoke the setting of a science fiction movie. But do not be fooled, in reality, it depicts the stunning southern hemisphere star trails over the ALMA antennas captured by a long-exposure image.

Bridging Borders

The European Southern Observatory is all about cooperation and bringing together different communities: scientists and engineers from fifteen countries; professional astronomers and educators; science communicators and the general public. All connected by their common interest in learning more about the Universe we inhabit.

The French journalist and photographer Serge Brunier first visited Cerro Paranal in May 1987. Long before there were any paved roads, and over three years before ESO Council selected the mountain as location for the Very Large Telescope. At the conical top of Paranal, there was room for little more than two small huts, a meteorology mast, and one or two pickup trucks. Brunier was deeply moved by the stark, alien beauty of the site and its surroundings, and by its scientific potential. Ever since, he has returned to Paranal repeatedly to catch this astronomical paradise on camera.

Serge Brunier is one of ESO's Photo Ambassadors. These are professional photographers who take their motorised camera mounts, digital equipment and fish-eye lenses to the photogenic sites in Chile where astronomers try to uncover the secrets of the cosmos. Some of his colleagues, such as Stéphane Guisard, Gianluca Lombardi and Gerhard Hüdepohl, are employed by ESO as astronomers or optical engineers. Others, like Babak Tafreshi, Christoph Malin or José Francisco Salgado, are independent science communicators like Serge. But all of them are captivated by the miraculous mix of desert, telescopes and the night sky, and they employ their photographic skills to bridge the border between professional astronomers and the general public.

Flags flying at Paranal
One of the foundations for ESO's success is the international collaboration. Here the flags of the ESO Member States are seen.

The story of the European Southern Observatory has always been a story of bridging borders. Fifty years ago, Jan Oort and Walter Baade mobilised astronomers from five European countries to work together on the construction of an observatory in the southern hemisphere — something that neither of those countries could have achieved on their own. The signing of the ESO Convention on 5 October 1962 by the five founding Member States — Belgium, France, Germany, the Netherlands and Sweden — marked the birth of what has become a success story of international cooperation.

Before long, other countries joined in. Denmark in 1967, Italy and Switzerland in 1982, Portugal in 2001. In 2002, the United Kingdom, which had withdrawn from the project-to-be in 1960, realised it couldn't afford *not* to be part of the most productive ground-based astronomical organisation in the world. Later, four more countries became full ESO Member States: Finland in 2004, Spain and the Czech Republic in 2007, and Austria in 2009. ESO offered them access to world-class astronomical facilities at the best possible locations. Meanwhile, their annual contributions offered ESO the possibility to further extend its ambitious science and instrumentation programme.

ESO's annual budget — some 130 million euros — is financed by the Member States, with each country's contribution being proportional to its net national income: Germany pays some 1750 times as much as the Czech Republic. Each Member State occupies two seats in the organisation's governing Council — usually one astronomer and one government representative — so all Member States have an equal say. In practice, however, ESO operates as a truly international organisation. Astronomy is a truly borderless science.

Serge Brunier talks about his first visit to Paranal in 1987

Vertigo. Glare. Dry, cold wind across the blue sky. Brown, beige and yellow Atacama soil beneath my feet. Not a cloud. Up there, a blinding Sun, unable to weaken the royal-blue sky. Above my head, a meteorology mast, whistling and vibrating with the gusts of wind. Just below, two tiny cabins and a generator. Francisco Gomez Cerda, in massive sunglasses, with brown skin and a dusty, thick moustache, wrapped in his jacket, is eager. He does not have many visitors, especially from so far away. Looking through a big logbook maintained since 1983, he turns the pages, and searches....

Francisco Gomez Cerda is a meteorologist who, with his two sons, tends the mountain. They are testing the sky for ESO at this unknown mountain. I came here on a bumpy track. At the top, there is room for only two barrack-style huts, one or two vehicles, and the meteorological mast. Install a telescope, here? Really, are you serious? The air is so dry that I can literally taste the high altitude. But no, 2660 metres — that's not so high.... Francisco Gomez Cerda recites statistics.... 1983, 1984, 1985, 1986, 1987, January, February, March.... Leaning over his shoulder, I read "photometric

A view of Paranal from 1987
When Serge Brunier first visited in May 1987, Cerro Paranal was shaped like a sugar loaf.

sky" in almost every line.... And then all of a sudden, Francisco rises, triumphantly. *"Ah, there, look! You see? It rained! There, a little over a year ago."*

I look around. Desert no matter where I turn. I've never seen anything like this. Thousands of hills and mountains: Cerro Cometa, Cerro Ventarones, Cerro Armazones, Cerro La Chira, Cerro Vicuna Mackenna. All located between 2600 metres and 3200 metres altitude, and rising, higher and higher, with their soft shapes and indistinct ochre, yellow and copper tones together, to the horizon. And then one point catches the attention in the midst of the sandy shades, and golden dust. The shining white pyramid of the Llullaillaco volcano stretches upwards to an incredible altitude. Isolated and incongruous, tiny and imposing at the same time. Dominating the Atacama Desert with its 6778 metres....

This was exactly 25 years ago, in May 1987. The supernova in the Large Magellanic Cloud still gleamed faintly

in the night sky, visible to the naked eye, and I had arrived in an astronomers' paradise. I could not remember the name of this mountain that ESO had invited me to visit. Francisco Gomez Cerda replied *"Cerro Paranal — we are at Cerro Paranal."*

Since 1987, hardly a year has gone by without me going back ... Some are obsessed with the ocean, or Sahara, or a mythical summit of the Alps or the Himalayas. I became a devotee of the Atacama Desert. Every time I go to Cerro Paranal, with its paved road, its levelled summit, its hi-tech Residencia, its one-hectare platform at the top, its busy control room, its four gigantic domes, and experience this feeling of being on an airfield for spaceships, I smile. Thinking back to the tiny windswept Chilean family who, for years, helped to discover the best astronomical site on the planet.

Recently, ESO's membership has even crossed the border between the northern and the southern hemispheres, when the Brazilian government signed the agreement for Brazil to become the fifteenth Member State, and the first non-European one. Currently the sixth largest economy in the world (and growing), Brazil will, after ratification of the agreement in parliament, greatly facilitate the development of ESO's next big project, the European Extremely Large Telescope. There's no reason why other countries, from whichever continent, might not follow in Brazil's footsteps.

Not only do Member States hitch a ride on a grand scientific voyage. It's good for their economies too. Although ESO doesn't apply an official "fair return" policy like its space science counterpart, the European Space Agency, most of the industrial contracts for the construction of new astronomical facilities are awarded to companies in the Member States, and it's perhaps not so surprising that the larger members often acquire the biggest contracts. For instance, for the Very Large Telescope, the 8.2-metre mirror blanks were cast by Schott in Mainz, Germany; the mirrors were polished by REOSC in France, and the telescope mounts and enclosures were built by the Italian industrial consortium AES.

But the smaller Member States are far from excluded economically. For instance, the Belgian company AMOS constructed Paranal's four 1.8-metre Auxiliary Telescopes, while the VLT interferometry delay lines were developed by Fokker Space and TNO–TPD (Netherlands Organization for Applied Scientific Research — Institute of Applied Physics in the Netherlands). What's more: many universities and scientific institutes all across Europe contribute to the development of ESO's science instruments — the sensitive "retinas" of the big telescopes in Chile. Without these high performance cameras and spectrographs, the VLT and its smaller siblings at La Silla would be blind.

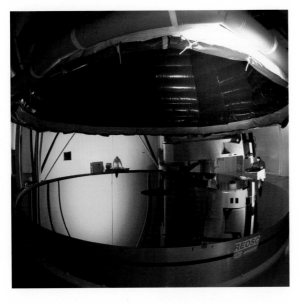

Last phase of polishing a VLT mirror at REOSC
Most of the industrial contracts for the construction of new ESO facilities are awarded to companies in the Member States. The VLT mirrors for instance were cast by Schott in Mainz, Germany, and polished at REOSC, France (seen here).

ESO pushes technology forward
Astronomer Rudi Albrecht in the Data Centre at ESO Headquarters in Garching bei München, Germany, which archives and distributes data from ESO's telescopes. He is in front of a rack containing a system with 40 processor cores, 138 terabytes of storage capacity and 83 gigabytes of RAM — over five million times more than the machine he used back in 1974.

Work on astronomy-related contracts can also bridge the borders with other fields of application. And while this process of technology transfer is of course not a main goal of ESO, it is a very welcome additional benefit for the Member States' industries, expanding and improving their experience at the forefront of technology, and opening up completely new markets. For instance, the development of wavefront sensors used in adaptive optics has found applications in eye surgery instruments, while the technology of optical fibres for the adaptive optics laser guide star system is now also used in telecommunications and medical equipment.

By organising workshops, seminars and summer schools, ESO offers a stimulating environment for promoting the proliferation of new technologies. And the list of success stories just keeps growing. Computer-controlled active optics, mirror blanks made of metal instead of glass ceramics, storage and mining of huge data volumes — in all these fields, industries and science consortia have gained experience that gave them an important edge in acquiring new, non-ESO contracts.

But for science and industry to play a pioneering role in developing ground-breaking instruments and technologies, there must be qualified people, bright inventors and visionary scientists to make that happen. That's why ESO also bridges the border from the lab to the classroom, by focussing on education. Today's children are tomorrow's engineers and astronomers.

Dental laser incorporating VLT technology
Technology transfer is a welcome additional benefit for the Member States' industries, expanding and improving their experience at the forefront of technology, and opening up completely new markets. LIMO Lissotschenko Mikrooptik produced some of the high-precision aspherical cylindrical lenses used in the VLT which led to a successful collaboration with the German dental company Oralia medical on the new dental diode laser seen here.

Unconventional visitors

"....this is really one of those once in a lifetime experiences. This is the best man has to offer at this time; a true pinnacle of technology and teamwork has been achieved in order to probe the vast reaches of our universe"

The Racing Green Endurance blog, 27 October 2010

Not surprisingly, Cerro Paranal, home to ESO's Very Large Telescope, has no train station or regular bus stop. The only way to get there — except by bicycle or helicopter — is by car. At the observatory, ESO operates a fleet of white cars and pickup trucks so encountering cars at Paranal is nothing special. Except, of course, when it's the latest Audi model, a brand-new Land Rover, or the luxurious BMW 6 series Grand Coupé. Over the past couple of years, all three car manufacturers have chosen Paranal as the preferred backdrop for advertising campaigns, usually because of the unique combination of stark natural beauty and high-tech buildings. Land Rover's campaign was even entitled "Perfect Places" — a description every ESO astronomer could agree to.

In October 2010, Paranal was also one of the pit stops for the Racing Green Endurance tour — a 26 000-kilometre trip from northern Alaska to Ushuaia along the Pan-American Highway taken by the SRZero electric sports car that had been developed by a student team from Imperial College, London. Maybe the most spectacular non-astronomical activity at Paranal was the filming of the closing scenes of Marc Forster's blockbuster movie *Quantum of Solace*, featuring Daniel Craig as secret agent James Bond and French actor Mathieu Almaric as the evil Dominic Greene. Paranal's Residencia served as Greene's hideout. But don't worry: the violent destruction of the place was filmed at a mock-up in a studio in London!

Languages on ESO's website

Languages on ESO's website
ESO reaches further than just the 15 Member States (blue) and the host nation Chile (purple). Anyone, regardless of nationality, can use the archived data from the many telescopes, and even request new observing time under certain conditions. ESO's website and press releases are also translated into many languages, indicated here in green and blue. Only the countries in grey are not yet covered with one or more of their official languages.

Astronomy is a thought-provoking endeavour, and all over the world, children just love planets, stars and galaxies. Both in Chile and in the ESO Member States, young students get in touch with the cosmos through special projects or school contests. The hope, of course, is that they will never lose interest again. Will Chilean student Jorssy Albanez Castilla from Chuquicamata, who, at the age of 17, suggested the Mapuche names for the VLT's Unit Telescopes, ever stop being excited about that achievement? And would winning a visit to Paranal, or — in some cases — the opportunity to carry out your own science programme with a large professional telescope, not make a lasting impression?

Even more important is reaching out to the public at large. ESO's telescopes are built with taxpayer's money, so the general public should be able to take part in the excitement. The education and Public Outreach Department makes that possible, by producing highly accessible leaflets, brochures and books. ESO also closely works together with planetariums and science museums in setting up astronomy exhibits and events.

Obviously, by far the most important platform for communication with the general public is the ESO website (www.eso.org) — a rich source of astronomical information, including thousands of beautiful pictures, a wealth of material on ESO's existing and future observatories, and even a well-stocked web shop. A trimmed-down version of the website is available in no less than 17 different languages. Especially popular is the ESOcast, hosted by Doctor J., aka Dr Joe Liske. This is a series of downloadable podcasts on the latest news and research from ESO, produced by the dedicated outreach team and loaded with breathtaking computer graphics, special effects, time-lapse videos and astronomical photography.

Hidden Treasures

In 2010, ESO invited amateur computer wizards and image processors to delve into the publicly available ESO Science Archive and to use real science data to create new and spectacular views of astronomical objects. Many results from professional astronomical detectors are never turned into colourful images, and the individual observations of a particular galaxy or nebula by a great variety of telescopes and instruments are not always combined into one single picture.

By late 2010, almost a hundred entries to the Hidden Treasures competition had been received. Working their way through many terabytes of data, people from all over the world had produced an impressive collection of beautiful photos, which were subsequently judged for their aesthetic and technical qualities.

Russian astronomy enthusiast Igor Chekalin won the first prize — including a trip to ESO's Very Large Telescope in Chile — with his stunning image of the dusty nebular complex Messier 78 in the constellation of Orion (seen here). Two other Chekalin images, one of a small group of galaxies and the other of the famous Orion Nebula, also ended up in the top twenty.

Hidden Treasures is one more way for ESO to bridge the borders between professional astronomy and the public at large, and to involve amateurs in the endeavour of revealing the beauty of the Universe we live in.

Doctor J.
Doctor J., aka ESO scientist
Dr Joe Liske has become
a popular figure represent-
ing ESO in ESOcasts and
documentary movies on TV.

Bridging the border between professional astronomy and the general public often involves print and broadcast media. Through its website and a dedicated media mailing list, ESO regularly issues press releases on new science results and technological milestones, usually accompanied by stunning visuals and broadcast-quality videos. As for imagery, there's also a Picture of the Week, portraying unique aspects of ESO's La Silla, Paranal and Chajnantor observatories, as well as the occasional soothing view of a colourful nebula or a majestic spiral galaxy. A qualified science outreach network, with representatives in every Member State and even in a number of non-member states, helps the local media to establish contacts with astronomers in the region, and to find a national angle to new developments and results.

Sometimes the Universe itself creates the opportunity for a big outreach event. On 8 June 2004, the dark silhouette of the planet Venus moved across the bright surface of the Sun, as seen from Earth — a rare cosmic spectacle that only happens twice every 120 years or so (the second

transit of the current pair occurred on 6 June 2012). ESO initiated a special programme aimed at teachers, students and the general public all across Europe. On transit day, the dedicated event website received more than 50 million hits in less than eight hours.

One thing is clear: in terms of outreach, nothing beats the Universe itself. Astronomy is a tremendously visual science, and photos of galaxies, star clusters and stellar nurseries fire our imagination. That's why a small part of ESO's telescope time is now dedicated to the Cosmic Gems Programme when weather conditions are not suitable for "real science". The idea: use instruments like the Wide Field Imager at the MPG/ESO 2.2-metre telescope at La Silla and VIMOS and FORS2 at the Very Large Telescope to obtain spectacular images just for education and public outreach — an initiative quite similar to the Hubble Heritage programme, carried out by NASA in the US.

And, of course, ESO's Photo Ambassadors, like Serge Brunier, greatly add to the public awareness of the European Southern Observatory. Their out-of-this-world images of the Atacama Desert and the Chajnantor Plateau provide a taste of the extreme terrestrial environments that ESO astronomers consider a second home. Their artistic portraits of astronomical observatories, telescopes and radio dishes — many of which are reproduced in this book — constitute the next best thing to actually travelling to La Silla, Paranal, or Chajnantor. And their awe-inspiring photographs of the Chilean night sky, of comets and planetary conjunctions, and of the Milky Way or the diffuse zodiacal light, reveal the splendour of the Universe that mankind tries to fathom.

When all borders are bridged and all barriers are breached, everyone can join the adventure. ESO offers a gateway to the cosmos. The Universe is yours to discover.

ESO Pictures of the Week
A selection of Pic-
tures of the Week from
the past few years.

The Running Chicken Nebula
ESO's Cosmic Gems programme observes interesting objects in the southern sky when the weather does not allow scientific observations. Hundreds of people from all over the world submitted their interpretations (right) of where the running chicken can be found in the nebula of the same name (left).

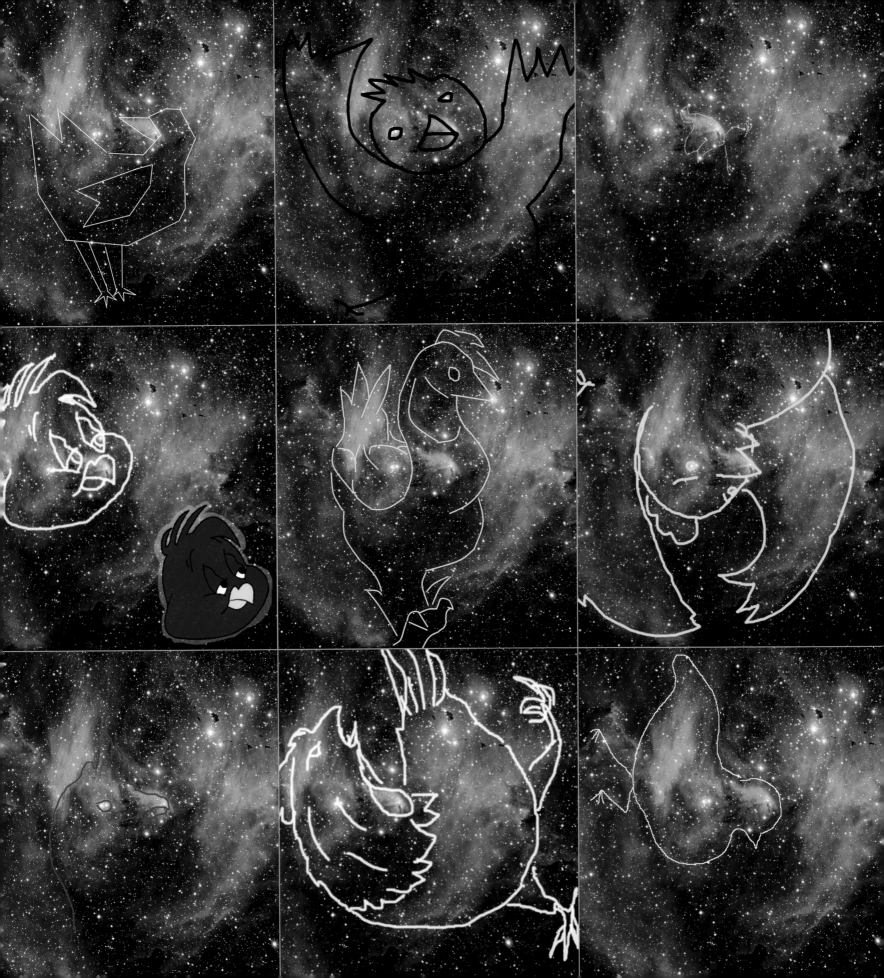

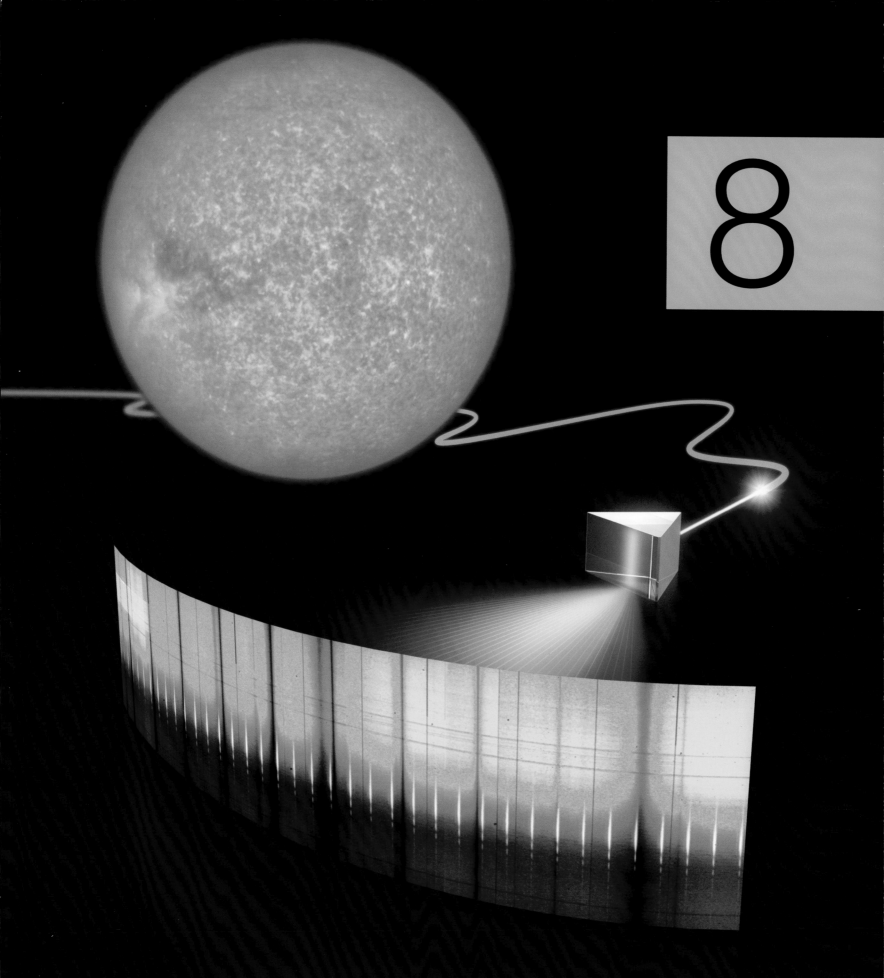

Catching the Light

Like hungry nestlings waiting to be fed, astronomical telescopes open up their mirrors to the night sky, to catch as many photons as they can. But dissecting starlight and wrenching out every possible bit of information about stars and galaxies is the work of high-tech cameras and spectrographs — the modern replacement of the human eye.

Artist's impression of a laser comb used in astronomy.
In the quest to invent ever more precise ways to dissect the light arriving to us from the cosmos, astronomers teamed up with quantum opticians to invent a way to use the new laser comb technique in astronomy. The laser comb has by now successfully been used to re-discover known exoplanets, and promises to be a powerful tool for finding new exoplanets, possibly as small as the Earth.

Imagine William Herschel walking into the control room of ESO's Very Large Telescope. This great 18th-century astronomer, discoverer of Uranus and surveyor of star clusters, nebulae and binary stars, would be delighted to learn about Neptune and the Kuiper Belt, the births and deaths of stars, the structure and dynamics of the Milky Way galaxy, and the existence of hundreds of billions of similar galaxies in the Universe. He would marvel at the dramatic spacecraft photos of Saturn's small moon Mimas, which he discovered, and at the captivating, colourful images of swirling gas clouds and colliding galaxies that we have become so accustomed to. But most of all, Herschel would be excited by the VLT's ability to catch infrared light from the depths of the cosmos.

In 1800, more or less by chance, Herschel discovered the existence of this long-wavelength radiation that our eyes cannot see. Infrared "light" — sometimes called heat radiation — is emitted by objects at room temperature or cooler. Dark and dusty clouds of cold molecular gas, invisible at optical wavelengths, can be imaged by infrared detectors. Moreover, these astronomical night goggles let scientists peer into the obscured cores of star-forming regions. They also reveal remote galaxies whose energetic radiation has been shifted to longer infrared wavelengths by the expansion of the Universe.

Today, we know that optical and infrared light are just two small sections of the full electromagnetic spectrum. In the second half of the twentieth century, astronomers have opened up many other wavelength regimes: radio waves, millimetre and submillimetre radiation, ultraviolet "light", X-rays, and the highly energetic gamma rays. Each part of the spectrum has its own story to tell, and to neglect one type of radiation is like attending a performance of Beethoven's Ninth Symphony with a hearing impairment that prevents you from hearing specific frequency bands.

Astronomers have devised sensitive electronic cameras to record all types of electromagnetic waves from space. Some of these waves can only be collected by big radio antennas, or by instruments on board Earth-orbiting space telescopes, but some near-infrared waves are caught and focussed by ground-based telescopes just like visible light. Just replace any optical camera with a sensitive infrared detector and make sure this detector is cooled enough so that it can record the feeble heat radiation from remote cosmic objects. Over the past few decades, electronic infrared detectors have become almost as sensitive as optical CCD cameras, and reveal the same level of detail.

Apart from cameras, astrophysicists use spectroscopic instruments to dissect starlight and to measure the distribution of energy over various wavelengths precisely. Spectroscopy reveals information on the temperature, motion and chemical make-up of stars, nebulae and galaxies, and it has become by far the most important tool in astronomy. Without spectroscopy, astronomers would just be staring at a beautiful landscape. With spectroscopy, they learn about the landscape's topography, geology, evolution and composition.

The optical and near-infrared cameras and spectrographs of present-day professional telescopes like the VLT are giant high-tech machines, each the size of a small car. Their purpose: to catch cosmic photons and recover every possible bit of information. If the Very Large Telescope is ESO's giant eye on the sky, these detectors constitute the eye's retina. The revolutionary findings that have made headlines over the past couple of years, in all possible fields of astronomy, would not have been possible without this versatile suite of instruments. One could even say that the focus on continuously developing new instruments for the VLT, built in collaboration with institutes in the Member States in return for observing time, has been fundamental for making the VLT the most advanced ground-based optical observatory in the world.

Collecting precious starlight
As soon as the Sun sets over the Chilean Atacama Desert, the VLT begins catching light from the far reaches of the Universe.

For instance, by virtue of their incredible sensitivity, ESO instruments have measured, what at the time of discovery was, the furthest quasar, the furthest gamma-ray burst, and the furthest galaxy. In all three cases, spectroscopy made it possible to determine the redshift of the faint object — the amount by which the emitted light has been stretched by the expansion of the Universe during its billion-year-long trip to Earth. Thus, Nial Tanvir of the University of Leicester, United Kingdom, used the ISAAC spectrometer and camera to reveal that the light from gamma-ray burst GRB 090423 took more than 13 billion years to reach Earth — it was emitted when the Universe was a mere 600 million years old. Instruments like ISAAC provide cosmologists with a view of the very early Universe.

HAWK-I
The HAWK-I instrument mounted on Yepun, Unit Telescope 4 of ESO's Very Large Telescope. HAWK-I covers about one tenth the area of the full Moon in a single exposure. It is uniquely suited to the discovery and study of faint objects, such as distant galaxies, young stars and planets.

The VIMOS spectrograph uses slit masks to catch the spectrum of about a thousand faint objects in one exposure. With VIMOS, a team led by Luigi Guzzo of the Brera Observatory, Italy, charted the three-dimensional distribution and the motions of huge numbers of remote galaxies. They show an intricate web-like pattern of clusters

The VLT reveals the Carina Nebula's hidden secrets This broad panorama of the Carina Nebula, a region of massive star formation in the southern skies, was taken in infrared light using the HAWK-I camera on ESO's Very Large Telescope.

and filaments, which can only be explained by assuming that empty space is filled with a mysterious dark energy that actually accelerates the expansion of the Universe. VIMOS also discovered a huge supercluster of galaxies at a distance of 6.7 billion light-years. Measuring 60 million light-years across and containing some ten thousand galaxies like the Milky Way, it is the largest known structure in the distant Universe.

NAOS-CONICA
With the help of adaptive optics the VLT instrument NAOS-CONICA performs a wide range of measurements: imaging, imaging polarimetry, coronography and spectroscopy.

To investigate the Milky Way galaxy, ESO astronomer Gayandhi De Silva used the ultraviolet spectrograph UVES to study the chemical makeup of stars in stellar clusters like the Hyades. She found that each cluster has its own distinct composition, reflecting the specific time and place of its origin. A similar window on the evolution of the Milky Way was obtained by Manuela Zoccali of the Catholic University of Chile in Santiago. Her UVES observations showed that the central Bulge of the Milky Way has a different chemistry from that of the flat disc, with larger relative amounts of oxygen. This suggests that the Bulge formed early and quickly in the galaxy's history, and is independent of the origin of the disc.

ESO's adaptive optics instruments NACO and SINFONI played a starring role in unveiling the supermassive black hole at the core of the Milky Way (see also p. 93). Reinhard Genzel and Stefan Gillessen of the Max Planck Institute for Extraterrestrial Physics in Garching, Germany led the effort to map the orbital motions of giant stars swirling around in the black hole's strong gravitational field. They also discovered a cool, elongated gas cloud three times as massive as the Earth that has been accelerated to a velocity of more than eight million kilometres per hour. In the summer of 2013, the cloud will pass close to the black hole's edge, and may well be shredded by its strong tidal forces.

VFTS 682

R136

Massive stars in the Large Magellanic Cloud
This view shows part of the very active star-forming region around the Tarantula Nebula in the Large Magellanic Cloud, a small neighbour of the Milky Way. At the upper left is the brilliant and very massive star VFTS 682 and at the lower right is the very rich star cluster R 136.

The power of present-day infrared detectors is evident from the breathtaking images captured by the HAWK-I camera since its installation in 2007. Because absorbing dust becomes transparent at infrared wavelengths, the instrument beautifully reveals the clean spiral structure of distant galaxies. And by stitching together hundreds of individual exposures, Thomas Preibisch of the University Observatory in Munich, Germany, created a stunning wide-field image of the Carina Nebula, one of the largest star-forming regions in our Milky Way galaxy. With its hawk-eye vision, the infrared camera shows hundreds of thousands of faint stars that have never before been imaged.

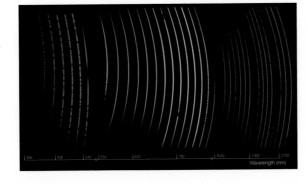

An X-shooter spectrum
This illustration shows the three spectra produced simultaneously by the efficient X-shooter instrument on ESO's Very Large Telescope. X-shooter can record the entire spectrum of a celestial object (in this example a distant lensed quasar) in one shot — from the ultraviolet to the near-infrared— with great sensitivity and spectral resolution.

An even more active stellar cradle, known as 30 Doradus, is located in the Large Magellanic Cloud, over 160 000 light-years away. Here, ESO's FLAMES spectrograph has found some of the most extreme stars known. For instance, VFTS 682, found by Joachim Bestenlehner of the Armagh Observatory in Northern Ireland, is 150 times more massive and three million times more luminous than the Sun. Surprisingly, while all other supermassive giant stars are members of tight stellar clusters, VFTS 682 is alone — a solitary heavyweight. Philip Dufton of Queen's University in Belfast, Northern Ireland, used FLAMES to measure the fastest-spinning normal star known — VFTS 102, with a rotational speed of over two million kilometres per hour. This supergiant may have been spun up like a top by mass transfer from a companion star that later underwent a supernova explosion.

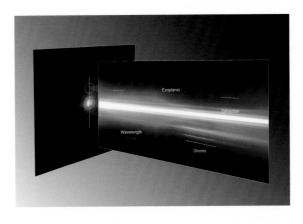

Spectrum of an exoplanet
The faint spectrum between the two lines near the top is that of a giant exoplanet, orbiting around the bright and very young star HR 8799, about 130 light-years away. This montage shows the image and the spectrum of the star and the planet as seen with the NACO adaptive optics instrument on the VLT. As the host star is several thousand times brighter than the planet, this is a remarkable achievement at the border of what is technically possible.

Obviously, huge telescopes like the VLT are also good at spotting cool, diminutive dwarf stars. A good case in point is CFBDSIR 1458+10B — the coolest star known to date, with a surface temperature of a mere 100 degrees Celsius. This "cup-of-tea-star" was detected by Michael Liu of the University of Hawaii in Honolulu, who exploited the powerful X-shooter spectrograph — a unique instrument that captures an object's spectrum in a very broad wavelength region, from the ultraviolet to the near infrared. X-shooter also found a mystery star almost completely consisting of hydrogen and helium, with 20 000 times fewer heavy elements than the Sun.

Using adaptive optics, Markus Janson from the University of Toronto, Canada, even succeeded in taking a direct spectrum of an extrasolar planet at a distance of 130 light-years. Weighing in at ten Jupiter masses, HR8799c has a cloud-top temperature of some 800 degrees Celsius. Frustratingly, the planet's spectrum doesn't seem to fit any popular theoretical models. Another indication that exoplanets are very different from what we're used to in the Solar System is the discovery, by Ignas Snellen of Leiden Observatory, the Netherlands, of a superstorm in the atmosphere of the gas giant HD 209458b. Using the VLT's CRIRES spectrograph, Snellen clocked the wind speed at an incredible 5000 to 10 000 kilometres per hour.

The detector revolution

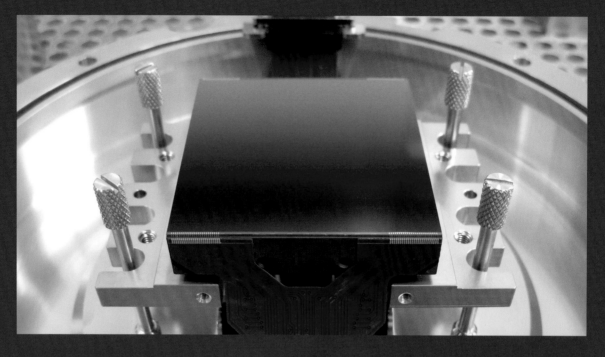

The large format MUSE detector
One of the 24 16-million pixel detectors to be used in the Multi Unit Spectroscopic Explorer (MUSE) second generation instrument for ESO's Very Large Telescope. MUSE is an innovative 3D spectrograph with a wide field of view, providing simultaneous spectra of numerous adjacent regions in the sky from 2013.

For centuries, astronomers relied on their eyesight and drawing skills to record what they observed through the eyepieces of their telescopes. It gave a nice artistic twist to the science of the Universe, but the human eye isn't very sensitive, and our mind has a tendency to see things that aren't really there.

The invention of photography in the first half of the 19th century brought about a massive improvement. Before long, astronomers were taking photos of the Moon, the planets, the stars and even of faint nebulae. Photographic plates are objective, they can be exposed for hours on end to catch more photons and record fainter objects, and the observations can be stored for later analysis.

But the real revolution came only in the 1970s, with the advent of electronic detectors known as charge coupled devices (CCDs). A CCD is much more efficient than a photographic plate, and the recorded data are available in digital format, for easy electronic processing and distribution. While the first primitive CCD detectors contained a mere 10 000 pixels or so, current electronic cameras have hundreds of millions of pixels, revealing breathtaking details.

The development of infrared detectors has been even more rapid and impressive. In the early 1970s, infrared radiation could only be measured using bolometer-like detectors, and infrared imaging was more like taking a "picture" with a coarse array of light meters. But today, infrared detectors are as powerful as optical CCDs, and the near-infrared images produced with ESO's VISTA camera are as detailed as optical photos.

As for spectroscopy, the biggest revolutions have come in increasing the observing efficiency. In the past, it could take hours to capture the spectrum of one galaxy. Using electronic detectors, this has been cut to a couple of minutes or so, and fibre optics make it possible to feed the light from a large number of objects simultaneously into the spectrograph, while integral field spectroscopy provides a means of taking a spectrum of every single pixel in the field of view.

CRIRES was also the instrument of choice for Emmanuel Lellouch of the Paris Observatory in France, who studies the tenuous atmospheres of cold, icy bodies in the outer reaches of the Solar System. Thanks to the instrument's high sensitivity, Lellouch was able to detect carbon monoxide and methane in the atmosphere of Neptune's big moon Triton, and to show that the atmospheric pressure has increased over the past twenty years because of seasonal warming. Similar measurements at the distant dwarf planet Pluto revealed that its thin nitrogen-rich atmosphere is 40 degrees warmer than its solidly frozen surface and contains unexpectedly large amounts of methane.

Still closer to home, Glenn Orton of NASA's Jet Propulsion Laboratory in Pasadena and Leigh Fletcher of Oxford University, United Kingdom, studied the giant planets Neptune, Saturn and Jupiter with the VISIR mid-infrared camera/spectrometer. In all three cases, the infrared measurements reveal temperatures and circulation patterns in the planet's atmosphere. In particular, the VISIR observations of the giant storm system in Saturn's northern hemisphere in 2011 complemented the close-up observations of NASA's Cassini probe orbiting the ringed planet. So, ground-based observations with large telescopes add context to spacecraft results.

And finally, ESO's FORS2 spectrograph discovered life in the Universe. Not on a remote exoplanet, but on Earth. By studying the detailed spectrum and the polarisation of Earthshine — the sunlight reflected by our home planet and subsequently reflected back by the night side of the Moon — ESO astronomer Michael Sterzik could deduce the existence of clouds, oceans and vegetation. In the future, the technique might well become an important tool to look for biosignatures on planets orbiting other stars than the Sun.

Instruments at the Paranal Observatory

Current instruments (as of July 2012)

Very Large Telescope

CRIRES
- Cryogenic high-resolution InfraRed Echelle Spectrograph.
- Spectral resolving power of up to 100 000 in the spectral range of 1–5 micrometres.

FORS2
- Visible and near-ultraviolet FOcal Reducer and low dispersion Spectrograph.
- Multi-mode instrument that can be used for imaging in the visible and for low-resolution spectroscopy.

FLAMES
- Fibre Large Array Multi Element ESpectrograph.
- Simultaneous study of hundreds of individual stars in nearby galaxies at medium-high spectral resolution.

UVES
- Ultraviolet and Visual Echelle Spectrograph.
- High-dispersion spectrograph, observing from 300–1100 nanometres, with a maximum spectral resolution of 110 000.

X-shooter
- Multi-wavelength (ultraviolet to near-infrared) medium-resolution spectrograph.
- The first of several second generation VLT instruments (see next page).

VIMOS
- VIsible MultiObject Spectrograph.
- Four-channel multiobject spectrograph and imager.
- Allows low-medium resolution spectroscopy of up to 1000 galaxies objects at a time.

ISAAC
- Infrared Spectrometer And Array Camera.
- Cryogenic infrared imager and spectrometer, observing in the 1–5 micrometre range.

VISIR
- VLT Imager and Spectrometer for mid-InfraRed.
- Diffraction-limited imaging at high sensitivity in two mid-infrared atmospheric windows (8–13 and 16.5–244.5 micrometres).

NACO
- NAos-COnica; NAOS is the Nasmyth Adaptive Optics System and CONICA is the name of a near-infrared imager and spectrograph.
- One of the instruments for the VLT's adaptive optics facility.
- High spatial resolution imaging of stellar systems and exoplanets

HAWK-I
- High Acuity Wide field K-band Imager.
- Near-infrared imager with a large field of view.

SINFONI
- Spectrograph for Integral Field Observations in the Near-Infrared.
- Near-infrared (1–2.5 micrometre) integral field spectrograph fed by an adaptive optics module.

VLT Interferometer

MIDI
- MID-infrared Interferometric instrument.
- An instrument used for both interferometric photometry and spectroscopy.

AMBER
- Near-infrared Astronomical Multi-BEam combiner.
- An instrument for photometric and spectroscopic studies from 1-2.5 micrometres, which combines the light of three telescopes, including all possible triplets of Unit Telescopes.

PRIMA
- Phase-Referenced Imaging and Micro-arcsecond Astrometry.
- A system designed to enable simultaneous interferometric observations of two objects, that are separated by up to 1 arcminute, without requiring a large continuous field of view.

PIONIER
- Precision Integrated Optics Near-infrared Imaging ExpeRiment.
- Visitor instrument for interferometry.
- Combines the light of four Unit Telescopes, or four Auxiliary Telescopes.
- Picks up details about 16 times finer than can be seen with one Unit Telescope.

VST

OmegaCAM
- Devoted to surveys.
- 32 CCD detectors that create images with a total of 268 megapixels, observing in the 0.3—1.0 micrometre range, field view of 1° x 1°.

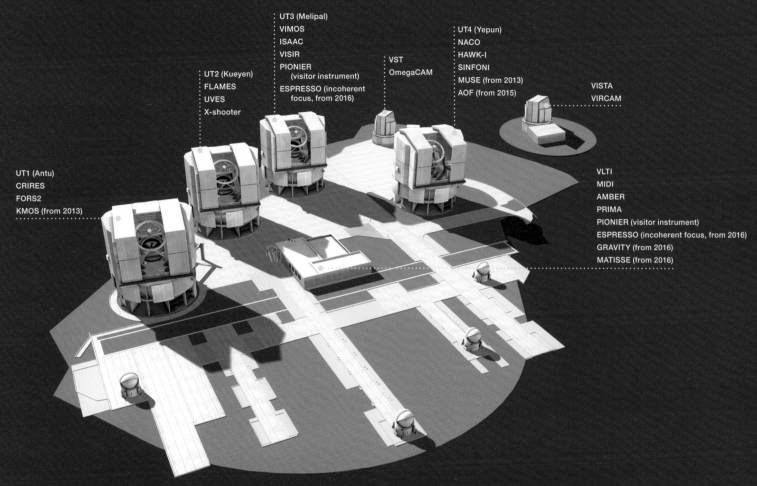

UT3 (Melipal)
VIMOS
ISAAC
VISIR
PIONIER
(visitor instrument)
ESPRESSO (incoherent
focus, from 2016)

UT4 (Yepun)
NACO
HAWK-I
SINFONI
MUSE (from 2013)
AOF (from 2015)

UT2 (Kueyen)
FLAMES
UVES
X-shooter

VST
OmegaCAM

VISTA
VIRCAM

UT1 (Antu)
CRIRES
FORS2
KMOS (from 2013)

VLTI
MIDI
AMBER
PRIMA
PIONIER (visitor instrument)
ESPRESSO (incoherent focus, from 2016)
GRAVITY (from 2016)
MATISSE (from 2016)

VISTA

VIRCAM
- Devoted to surveys.
- 16 special detectors sensitive to infrared light, with a combined total of 67 million pixels, observing in the 0.84—2.5 micrometres range, field of view of 1° × 1.5°.

Future second generation instruments on the VLT

KMOS (from 2012)
- K-band Multi-Object Spectrometer.
- Cryogenic infrared multi-object spectrometer with 24 robotic arms to pick off objects in the field of view.

MUSE (from 2013)
- Multi Unit Spectroscopic Explorer.
- A huge "3-dimensional" spectrograph that provides a spectrum for each pixel.
- Will provide complete visible spectra of all objects contained in "pencil beams" through the Universe.

SPHERE (from 2013)
- Spectro-Polarimetric High-contrast Exoplanet REsearch instrument.
- High-contrast adaptive optics system to study planets around other stars.

AOF (from 2015)
- Adaptive Optics Facility.
- Converts UT4 into an adaptive telescope with four sodium lasers and a deformable secondary mirror.

ESPRESSO (from 2016)
- Echelle Spectrograph for Rocky Exoplanet- and Stable Spectroscopic Observations.
- High-resolution, fibre-fed and cross-dispersed echelle spectrograph for the visible wavelength range.
- Can be used by any Unit Telescope or all four together to detect planets orbiting other stars.

GRAVITY (from 2016)
- General Relativity Analysis via VLt InTerferometrY.
- Four way beam combination instrument for interferometry.
- Uses all four Unit Telescopes to measure precise angular distances between objects and also perform imaging.

MATISSE (from 2016)
- Multi AperTure mid-Infrared SpectroScopic Experiment.
- Image reconstruction instrument for interferometry.

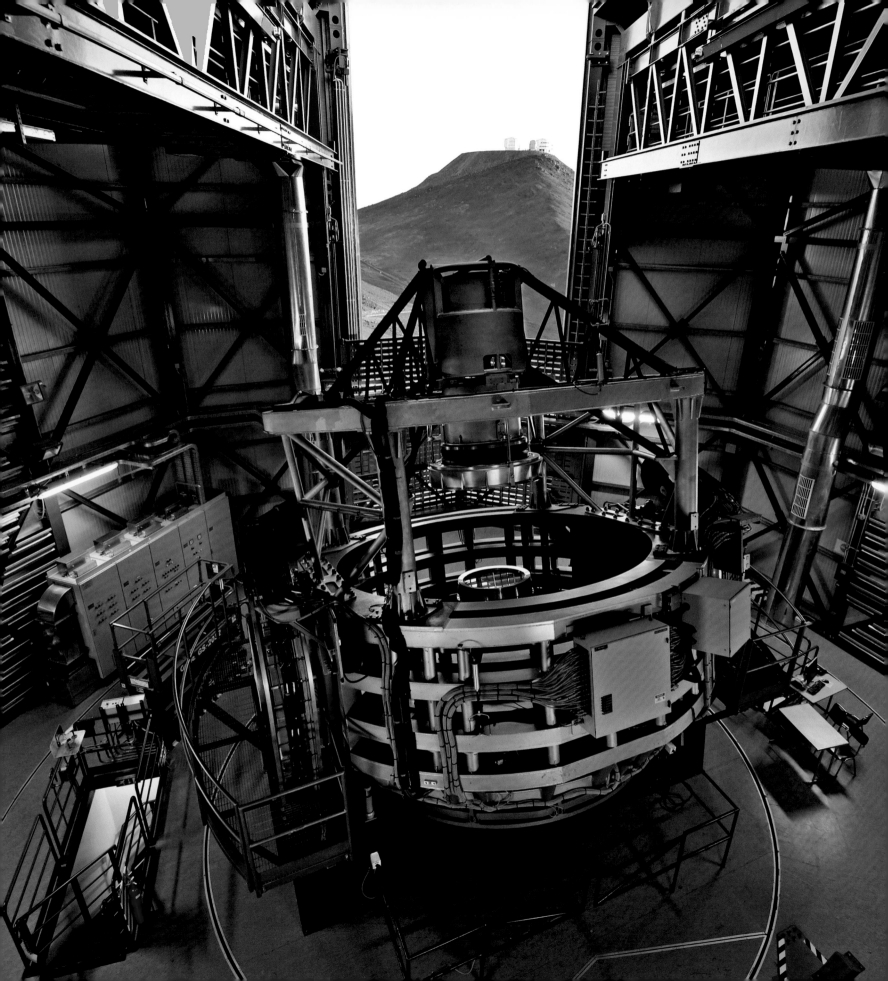

Apart from the big cameras and spectrographs that are mounted behind the four Unit Telescopes, the VLT also features two high-precision instruments for interferometric observations: AMBER and MIDI. Among other things, they have revealed puffs of gas surrounding both red giants and stars-in-the-making, and the sizes and shapes of asteroids.

And of course, the telescopes at ESO's La Silla Observatory each sport their own suite of instruments: an infrared camera and a spectrograph on the New Technology Telescope; the Wide-Field Imager on the MPG/ESO 2.2-metre telescope, and the powerful HARPS spectrograph on the 3.6-metre telescope, which is the most prolific ground-based exoplanet finder.

For ALMA the instrumentation is a very special story. Each ALMA antenna contains a set of state-of-the-art receivers — sensitive detectors which measure the incoming radiation in different millimetre and submillimetre wavelength ranges. To achieve their extreme sensitivity, they are cooled to a temperature of just -269 degrees Celsius using helium gas in a closed-cycle cryocooler (an advanced cooling device similar to that in a refrigerator). As with the rest of ALMA the receivers are developed in a close-knit partnership, and produced by institutes in Europe, North America, and Japan.

Two telescopes at Paranal, both with giant electronic cameras, deserve a special mention. At a small peak just a few kilometres from Cerro Paranal sits the 4.1-metre Visible and Infrared Survey Telescope for Astronomy (VISTA), equipped with a 3-tonne, 67-million-pixel camera. VISTA, in operation since late 2009, was developed by British universities as part of the United Kingdom's entrance fee to ESO. Meanwhile, at the main VLT platform, ESO and the Italian National Institute for Astrophysics built the 2.6-metre VLT Survey Telescope (VST), with its 268-million-pixel OmegaCAM imager built by a Dutch-German-ESO consortium. These are among the largest astronomical survey telescopes in the world. They carry out dedicated surveys of large swaths of the sky, and interesting objects

discovered by the two telescopes can be studied in much more detail by the Very Large Telescope.

In William Herschel's day, just over two hundred years ago, all of astronomy was done by actually peering through the eyepiece of a telescope. A century ago, photographic glass plates and crude spectroscopes had replaced the imperfect human eye. But only in the past couple of decades have high-tech electronic cameras and detectors started to reveal the intricacies of the Universe that we live in. And the end is not yet in sight. At ESO, and at universities and scientific institutions all across Europe, astronomers, engineers and opticians are developing a whole new generation of even more powerful instruments, both for the Very Large Telescope and for the future European Extremely Large Telescope. Without doubt, ESO's observatories in Chile will lead the way for a long time to come.

VISTA at sunset
This spectacular view of the VISTA telescope was taken from the ceiling of the building during the opening of the enclosure at sunset. VISTA is the largest survey telescope in the world and it is dedicated to mapping the sky at near-infrared wavelengths.

The VLT Survey Telescope
The VLT Survey Telescope is the latest telescope to be added to the Paranal Observatory. It is the largest telescope in the world designed for visible-light sky surveys.

Director General Riccardo Giacconi

Name: Riccardo Giacconi
Year of Birth: 1931
Nationality: Italian
Period as Director General: 1993–1999

What makes ESO special?
ESO is a great place in which to do science. The ESO Council gives the Director General the freedom to utilise the modern management techniques best suited for the execution of complex and technically challenging projects. Transparency and communication both in the vertical and horizontal direction can be established, permitting a shared vision of the goals of the organisation by the entire staff. Major scientific and technical decisions can be made in a cooperative mode, while allowing individuals full freedom and responsibility for execution. Scientific staff initiatives and creativity can be encouraged. Tight scheduling and spending discipline can be maintained, with no micromanagement. Personnel policies based on merit and recognition of individual accomplishment can be used to motivate the staff. Relations between ESO and national research institutions, as well as industry can be established on sound scientific and technical basis rather than being dictated by politics. I believe that the support of the ESO Council for these policies was essential to the success of the Very Large Telescope, but they represent a departure from those found in many other research institutions.

What was the greatest challenge during your time as ESO's Director General?
The greatest challenge during my term was the establishment of amicable and cooperative relations with the Chilean astronomical community and the Chilean government. Given the ambitious plans for the Paranal Observatory and the construction of the Atacama Large Millimeter/submillimeter Array, it was essential to resolve longstanding problems of relations with the local labour force, with the recognition of Chilean contributions over the years and the creation of favourable conditions in which to carry out the work. This process took several years, but led to a very constructive relationship.

How do you see ESO's future?
I have great confidence in the future of ESO. Astronomy is currently in the position of posing the great scientific questions for physics on the nature and composition of the Universe. Technological developments in all branches of observational astronomy will permit giant new steps in knowledge. The scale of the enterprise necessary to design, construct and operate these new facilities goes well beyond national capabilities. In the last fifty years, Europe has established ESO as an organisation capable of providing the necessary technical and management competence to do well in the worldwide competition for the next generation.

Sunset at Paranal Observatory
With such excellent observational conditions as seen here, it is no wonder that ESO has placed two of their most precious telescopes on Cerro Paranal: the Very Large Telescope (VLT, back) and the Visible and Infrared Survey Telescope for Astronomy (VISTA, front).

The Desert Country

The astronomical facilities of the European Southern Observatory are located in some of the most remote and hostile environments on Earth. Working and living in the Atacama Desert — the driest place on the planet — is a challenge in many different ways. But despite all these hardships, Chile feels like a second home to ESO's scientists and engineers.

With its pitch-black skies and cloud-free, bone-dry climate, northern Chile is an astronomers' paradise. Yes, Mauna Kea on Hawaii and La Palma in the Canary Islands offer excellent observing conditions. And yes, there's a lot of astronomical activity going on in the American Southwest and in South Africa. But nothing beats Chile. The country accommodates the vast majority of the world's largest telescopes, including the European Southern Observatory's Very Large Telescope at Cerro Paranal and the international ALMA Observatory at the Llano de Chajnantor. ESO's future 39.3-metre European Extremely Large Telescope will also be constructed here, at Cerro Armazones. Chile has become the home of European astronomy.

The Republic of Chile is the longest country in the world. It stretches over 4300 kilometres, between southern latitudes 17° and 56°, covering an area of over 750 000 square kilometres. With more than 17 million inhabitants, one third of them in the capital city of Santiago, Chile is *per capita* Latin America's most prosperous country, with a gross domestic product of some 225 billion euros, and a record of steady economic growth. Despite a chequered political history, it is now a stable democracy, currently headed by President Sebastián Piñera Echenique.

Vicuñas in the Atacama
Desert in Chile's Region II
This photo of five vicuñas
was taken not far from
ALMA and APEX. The
vicuña is one of two wild
South American camelids.
It is a relative of the llama.

One of the many faces of the Atacama Desert
The picturesque Laguna Miscanti in the Los Flamencos National Reserve in the Atacama Desert.

Perú

Bolivia

Pacific Ocean

CHILE

Argentina

Atlantic Ocean

Map of mainland Chile
The Republic of Chile is
the longest country in the
world. The area marked
contains all ESO's obser-
vatory sites (see p. 28 for
an enlarged version).

Chile's first settlers arrived some 10 000 years ago.
The Native Indian territories were invaded by Spanish
conquistadores in 1536. The country regained its independence in 1810, but became heavily involved in the War
of the Pacific with Peru and Bolivia (1879–1883), and in the
course of the 20th century made a transition from a parliamentary republic to a presidential system that remains to
date. Soon after ESO decided to establish its observatory
in Chile, in late 1963, the republic started to see important economic and political reforms, under presidents
Eduardo Frei Montalva and Salvador Allende Gossens.
Relations with the Chilean government became much
tenser in the years following the military coup by Augusto
Pinochet Ugarte in September 1973. Democracy was fully
restored in 1990.

Harbouring American and European observatories has
been a big boon to Chilean astronomy. Up to ten percent
of the observing time on all facilities is reserved for meri-
torious proposals by Chilean astronomers. In response,
national universities like the Universidad de Chile, the Pon-
tificia Universidad Católica de Chile (both in Santiago) and
the Universidad de Concepción started to offer Masters
and PhD programmes in astronomy. Government funding
of astronomy is now strong, and the number of astronomy
graduate students amounts to some 200. Universities also
cooperate in the Santiago-based Centre for Excellence in
Astronomy and Associated Technologies (CATA), and they
successfully attract scientists from abroad.

**The Chilean night sky at
Paranal**
This outstanding image
of the sky over two Auxiliary Telescopes at Paranal
depicts several deep-sky
objects. Most notable is
the Carina Nebula, which
stands out in the mid-
dle of the image, glowing
in an intense red, a sign of
ongoing stellar formation.

Life at Paranal

The kitchen in the Residencia
Preparing all meals for the staff, contractors and visitors at Paranal is no small task.

The Paranal base camp, with its unique Residencia, lies 130 kilometres south of the Chilean harbour town of Antofagasta, at an altitude of 2360 metres, on the slopes of Cerro Paranal, which is home to ESO's Very Large Telescope. Here, in the heart of the driest and probably oldest desert in the world, lies a complete village with a hotel, office space, laboratories, warehouses and technical buildings, including a huge plant to re-aluminise the giant 8.2-metre mirrors of the VLT's four Unit Telescopes.

The VLT employs over 170 FTEs (full time equivalents) at the Paranal Observatory, and on average, the base camp's population amounts to some 120 people, all of whom have to cope with the harsh conditions of the desert, like excessive sunshine, extremely low humidity (which occasionally drops below 5 percent) and low oxygen levels because of the elevation. Luckily, the Residencia offers lush and relaxing conditions, and includes a swimming pool and a sauna. Also, Paranal paramedics have a fully equipped ambulance to take people to the nearest hospital in case of emergencies.

All electrical power has to be generated on site, by a 2.4-megawatt multi-fuel generator. Moreover, all the water that is consumed and used at Paranal has to be trucked in from Antofagasta: some 60 000 litres per day. To feed the hungry scientists and engineers the Residencia's cafeteria uses over 12 000 kilograms of flour and over 80 000 eggs per year.

Astronomy is a challenging science, but observatory logistics poses its own problems!

Little wonder, given the excellent observing conditions. For thousands of years, Chile's original inhabitants must have marvelled at the night sky, although hardly any records remain, the oldest being a report of the observation of a lunar eclipse in June 1582 by Spanish soldier Pedro Cuadrado Chavino. Today, thanks to the sparsely populated rural areas, the sky is almost as dark as it was in the days before electric lighting. Northern Chile offers more than 315 cloudless nights per year, with a remarkably dry and steady atmosphere — the most important ingredients for a favourable astronomical climate. Only Antarctica could offer better conditions, especially for infrared and submillimetre astronomy, but in terms of logistics and infrastructure, Chile wins hands down.

Astronomers have to thank geology for creating their earthly paradise. Stretching over more than 100 000 square kilometres, the Atacama Desert is the driest and possibly the oldest desert on the Earth's surface. The cold Humboldt current in the Pacific Ocean creates a low inversion layer, preventing moist ocean air from crossing the Cordillera de la Costa, the coastal mountain range that borders the desert on the west, while convective rain clouds from the Amazon Basin are blocked by the towering Andes range to the east. As a result, the Atacama has been bone-dry for millions of years.

The average precipitation is about one millimetre per year. Some places haven't seen a drop of rain for many centuries. The period between 1570 and 1970 was exceptionally dry. As a result, the landscape has an eerie, Mars-like appearance, with sand dunes and boulder-strewn valleys. Vegetation is almost non-existent; animals, birds and insects are rare. Only the hardiest micro-organisms are able to survive in this arid environment. It should come as no surprise that the Atacama Desert has been used as the backdrop for science fiction movies, as a test bed for Mars rovers and as a natural laboratory for research on extremophile bacteria.

By far the highest humidity level in the entire Atacama Desert can be found in the Paranal tropical garden, part of the impressive Residencia, which also boasts its own swimming pool. Obviously, providing ESO's observatories with water is one of the big challenges of working in the driest place on Earth. At La Silla, groundwater is pumped up from the Pelícano base camp, and at ALMA, close to the oasis of San Pedro de Atacama, water wells have been dug at the Operations Support Facility at 2900 metres altitude. But at Paranal, every drop of water has to be trucked in from Antofagasta, 130 kilometres to the north. Three times a day, a water truck resupplies the million-litre-capacity observatory water tanks.

Water for Paranal
One of three daily trucks that deliver water to Paranal.

Supplying power is another challenge. Paranal and nearby Armazones may hook up to the national power grid in the future, but for now, the observatory has its own generators. At Paranal, a 2.4-megawatt multi-fuel generator and three smaller back-up generators provide power for the Very Large Telescope, while a permanent seven-megawatt power plant has recently been installed at the ALMA OSF. Surprisingly, solar (or wind) energy turns out not to be a viable alternative, mainly because there's no easy way to store power if you're not connected to the grid. At most, the energy needed to chill the VLT enclosures could be provided by the Sun.

ESO Representative Massimo Tarenghi

Name: Massimo Tarenghi
Year of Birth: 1945
Nationality: Italian
With ESO since: 1973 (as astronomer, and then as project or programme manager for many large telescopes, including the MPG/ESO 2.2-metre telescope, NTT, VLT and ALMA; director of Paranal Observatory and of ALMA, and ESO Representative in Chile)

What makes ESO special?

ESO was set up to do great things — to forecast the future of astronomy and to turn dreams of astronomers into reality. This happened with the 3.6-metre telescope, the NTT, the VLT, and ALMA. And now we have the great step of the E-ELT. These were created through the spirit of European cooperation.

What was the greatest challenge during your ESO career?

To put together different cultures and languages in the different projects and to help change the organisation as time went by and projects evolved in complexity. In a sense this was almost like going from a family-run business to an international corporate structure. It was also challenging to interact with big industrial companies and consortia, and create the enthusiasm needed to build ground-breaking prototypes for the future of astronomy at the extreme limit. They needed to be convinced that the research and development needed to break the barriers in science also would be beneficial for them financially.

What is your favourite ESO anecdote?

The most spectacular for me was the first light of the NTT. We had a near-disaster just a few days before. The shape of the primary and secondary mirrors did not "match" each other — somewhat similar to what happened to the Hubble Space Telescope. ESO optical engineer Lothar Noethe managed in two days to modify the mirror shape in the necessary way with the actuators for the active optics. We did the first light observations from the Headquarters in Garching remotely and obtained sharper images than had ever been done from the ground. This was a live demonstration that the active optics and dome worked, and a new way of doing astronomy from the ground was born — this was the Galileo moment of my life!

How do you see ESO's future?

I am convinced that ESO in 50 years will continue to have dreamers inside and outside the organisation suggesting new ideas for how to go deeper and sharper in the exploration of the Universe. ESO does not just operate observatories. ESO plans and implements new facilities and can do long-term planning on 20-year timescales. This is one of our great strengths.

**Conjunction
over Paranal**
Guillaume Blanchard,
an ESO optical engi-
neer, is photograph-
ing a planetary con-
junction with the Moon
from the VISTA site.

Working at high altitude
Oxygen, sunscreen and polar clothing are prerequisites for working at the ALMA Operations Site at 5000 metres altitude.

Paranal mechanical engineer Juan Carlos Palacio
This is not a *minero* from an Atacama mine, but mechanical engineer Juan Carlos Palacio at work at the VLT.

Sunburn and dehydration are serious issues for people working in the desert, but the elevation of the ESO sites poses severe additional physical risks. Altitude sickness is a particular problem at the ALMA Array Operations Site (AOS), at 5000 metres altitude, where low oxygen levels may induce headaches, dizziness, and nausea. Visitors and astronomers alike are also urged to consult the observatory paramedic if they are "seeing stars"! Brief medical examinations are required of everyone visiting the ALMA high site, and bottled oxygen is provided on demand, even though the control room at the AOS is pressurised. At La Silla (2400 metres) and Paranal (2635 metres), altitude sickness is less of a problem, but even there the low air pressure can cause breathing problems, fatigue and loss of concentration.

Of course, the original inhabitants of the Atacama and the altiplano are much better equipped to cope with the effects of high altitude, whether by genetic adaptation or by chewing coca leaves. Long before astronomers set their sights on the Llano de Chajnantor, the volcanic area was littered with sulphur mines. Mining has been Chile's main source of income for centuries, with the Chuquicamata mine north of Calama — the world's largest open mine — producing almost a million tons of copper per year. In Chile, mining completely overshadows astronomy, both in terms of the scale of the infrastructure and of the environmental impact.

While blowing off the top of a mountain to make room for an observatory may seem to be just as disrespectful to the landscape as digging a kilometre-deep hole like the Chuquicamata mine, ESO is taking the utmost care to protect the delicate environment, especially in the Chajnantor area, which is home to the giant cactus *Echinopsis atacamensis* (aka the Cardón grande). A number of prime specimens, some six metres tall and weighing several tons, were carefully relocated during the construction of the ALMA access road. Another important point of concern is the preservation of prehistoric sites and artefacts, like the El Molle petroglyphs close to La Silla and the remains of 2000-year-old settlements in the San Pedro area near ALMA.

Atacama impressions

Desierto florido
In years with unsual amounts of rain, the barren, dusty desert landscape can transform in a multi-coloured tapestry of blooming desert flowers, in an event known as *desierto florido*.

Yesterday for a moment I imagined myself seeing this place for the first time and it impacted me. Petrified lava flows, the volcanoes and enormous domes. The Atacama Salt Flat basin with Kimal in the background. The colored hills, unimaginable vegetation, animals and birds, and a town built deep in the bottom of a gorge.

Unique Atacama.

Dirt, rocks, sun, salt, silica, iron, sulphur, lithium, gypsum, … lots of minerals … water, life, flesh and bones. The Salt Range, salt flats, gorges, the Altiplano, volcanoes. Everything under a sky that commands your attention.

Here we are lucky to be surrounded by such beauty. Hidden beauty that gives you all the years you need to see and understand. You can watch the setting sun behind the Domeyko mountain range over and over again. The colors change on the mountains (browns, reds, purples) but you discover that they stand unmoving. The black shadows accelerate quickly at dusk, faster than what it seems.

Dust everywhere and dry hands.

I am waiting. This evening the full moon rises behind the salt flat, a magic event that hides from us more than 3,000 stars that can be seen in this dark sky.

I move to the shade but I freeze.

Over everything I can hear the silence.

In the end it becomes real.

Excerpts from the photo book *Atacama Incalculado* by Maurice Dides Nazar (www.mauricedides.cl)

Situated in the very heart of the Atacama Desert, San Pedro is a laidback little town with adobe houses, dusty streets and sleepy dogs. From here, tourists visit the area's spectacular natural wonders, including the surreal Valle de la Luna, the endless Salar de Atacama, and the El Tatio geyser fields. While the area around ESO's Paranal Observatory is almost devoid of every form of life, the San Pedro region is home to flamingos and llamas, to rabbit-like viscachas and ostrich-like rheas, and to the elegant vicuñas that populate the altiplano.

Once every four or five years, the barren, dusty landscape further south in the Vallenar–Copiapó area transforms in a multi-coloured tapestry of blooming desert flowers, in an event known as *desierto florido*. The Atacama never ceases to surprise and impress, as many writers and pho-tographers will testify.

Some of the Atacama's surprises, however, are less benign than others. Just south of Chajnantor is the most active volcano of the central Andes range: Láscar. This stratovolcano, with a summit elevation of 5592 metres, regularly produces steam and ash clouds, as well as large eruptions, like the ones in 1993 and 2000. The volcanic activity of the region is the result of the Nazca Plate diving under the South American Plate — a tectonic process that also produces numerous earthquakes. The great Chil-ean quake of 27 February 2010 was the seventh biggest earthquake ever recorded by seismographs. But, apart from a power outage at La Silla, the quake did not cause any damage at the ESO observatories. At the VLT the 23-tonne mirrors of the four Unit Telescopes are auto-matically clamped by mechanical safety supports should there be any critical seismic activity.

Accidents do happen
Working in the desert more than 10 000 kilometres from Europe involves challenges, risks and dangers. Despite a major setback for the VST project when the main mirror was shattered on the way to Chile in 2002, the telescope had a very successful first light.

But accidents do happen. The desert is a hostile and dangerous place, and large high-tech operations are never without risk. In late 1986, reflected sunlight caused a fire in the secondary optics of the Swedish-ESO Submillimetre Telescope, which was under construction at the time, and six years later, asphalt work on the roof of ESO's 1-metre telescope at La Silla also started a fire. Both at Paranal and at Chajnantor, driving accidents have unfortunately taken their toll, and all sorts of quirky technical problems have led to delays in the development of telescope instrumentation.

By far the biggest technological setback was the complete destruction of the fully polished 2.6-metre primary mirror of the VLT Survey Telescope as it was being transported by sea in May 2002. On arrival in Antofagasta, it was found to be completely shattered. A replacement mirror took four years to cast, grind and polish. The transport of the new mirror encountered its own misfortune in 2009: due to severe leakage of sea water, the delicate mirror cell was damaged, and the whole cargo had to be shipped back to Europe again for time-consuming repairs. You can imagine the sighs of relief from scientists and engineers when the VST finally produced its first dramatic astronomical images in June 2011!

Living and working in the desert involves challenges, risks and dangers — not only in terms of the technology, but also on a human level, when multi-year job assignments in Chile put a strain on personal well-being, relationships and family ties. But all these hardships are very much worth the effort. Taking part in the grand cosmic adventure of exploring the Universe, being a member of the international ESO community, and enjoying the intense beauty of the Atacama Desert is a privilege that provides an unforgettable experience. With the construction of ALMA well under way, and the European Extremely Large Telescope approved, the future has a wealth of new unforgettable experiences in store for ESO's employees, contractors and their families.

Sharing a dream
The culture, background and language of people working for ESO is very different, but they are all part of the ESO dream — to connect humankind with the cosmos.

Giant Eye on the Sky

Four centuries after Italian physicist Galileo Galilei trained his first small telescope on the heavens, European astronomers are starting to build the biggest optical-infrared telescope in the history of mankind. With the future 39.3-metre European Extremely Large Telescope, Europe takes the concept of "reaching for the stars" to a completely new level.

Rendering of the E-ELT
This artist's rendering shows the E-ELT at work, with its dome open and its record-setting 40-metre-class primary mirror pointed to the sky.

Miraculously, the shock absorbers of the jeep don't break during the trip to Cerro Armazones. Even more miraculously, our kidneys survive. The unpaved road, right through the barren heart of the Atacama Desert, is a jumble of rocks and potholes, and our driver, German engineer Volker Heinz, seems to enjoy our sufferings — he is driving at breakneck speed. About an hour after leaving Paranal's Residencia, our small group arrives at the conical summit of Cerro Armazones, 3060 metres above sea level. The view is magnificent, with the Very Large Telescope silhouetted against the western sky. And we feel privileged to visit the future site of the largest telescope in the history of mankind.

For the full four centuries since its invention in 1608, the history of the telescope has been a race for ever bigger lenses and mirrors. Larger telescopes catch more starlight, reveal fainter objects and finer detail, and reach out further into the cosmic depths of space and time. Fifty years ago, it looked as if a natural limit had been reached with the 5-metre Hale reflector at Palomar Mountain, but in the 1980s, new technologies like thin mirrors and active optics enabled the construction of 8- and 10-metre-class instruments, including the European Southern Observatory's Very Large Telescope at Paranal. So why not take another step forward? What's wrong with *really* big telescopes?

The European Extremely Large Telescope (E-ELT) is not just really big — it's monstrous. Sporting a primary mirror almost 40 metres across — half the size of a soccer pitch — it will collect more starlight than all existing 8- to 10-metre-class telescopes combined. For many decades, ESO's future workhorse facility will be the world's biggest eye on the sky. Primordial galaxies at the edge of the observable Universe, stellar nurseries and mysterious black holes in the Milky Way, frozen bodies on the outskirts of our Solar System and even Earth-like exoplanets that might harbour life — nothing will escape the E-ELT's eagle-eyed view. We are on the verge of a new revolution in the history of astronomy.

Large telescope mirrors are delicate products, polished into a perfect parabolic shape with a precision of a few nanometres over an area of tens of square metres. In practice, it's impossible to make single-piece, "monolithic" mirrors larger than just over 8 metres across — a limit partly set by the need to transport and handle them. But by combining several individual mirrors in one telescope, it's possible to catch the same amount of starlight and to reach the same sensitivity and resolution as a much larger mirror. For instance, the Large Binocular Telescope in Arizona has two 8.4-metre mirrors, providing the same light-collecting power as an imaginary 11.8-metre telescope.

Combining mirrors to build monster telescopes is the trick used by American, Australian and Korean astronomers in the construction of the Giant Magellan Telescope (GMT) at Cerro Las Campanas in Chile, just north of ESO's La Silla Observatory. Funding permitting, the GMT will consist of seven 8.4-metre mirrors on a single mount, yielding the same performance as a single 24.5-metre telescope. This approach is a bit like pulling a massive load with seven trucks if there's no practical way to construct one super-heavy hauler.

But there's another option, comparable to pulling the load, not with seven trucks, but with hundreds of individual weight-lifters. For instance, the 10-metre primary mirrors of the twin Keck telescopes at Mauna Kea, Hawaii, and the Gran Telescopio Canarias at La Palma are composed of 36 hexagonal segments, and there's no reason why this approach can't be extended beyond 10 metres. Indeed, the design for the Thirty Meter Telescope (TMT), planned for construction at Mauna Kea, calls for 492 1.45-metre segments, with a total surface area of over 600 square metres — nine times as much as the collecting area of a 10-metre telescope.

To ESO staff members Roberto Gilmozzi and Philippe Dierickx, it was clear from the start that the segmented mirror approach would be the preferred way to construct a monster telescope. Back in 1999 — the same year as the inauguration of the Very Large Telescope — they came up with a preliminary design for a telescope with an unbelievably large, 100-metre mirror, consisting of over 3000 segments. Gilmozzi and Dierickx were unable to identify any serious technical show-stoppers for their OverWhelmingly Large Telescope. They even believed the design could be scaled up to 130 metres.

Of course, money and practical feasibility are different matters altogether. While a detailed OWL concept study did a great job in exploring new ways and technologies for building monster telescopes, it soon became clear that 42 metres was a more realistic figure — a mirror with twice the collecting area of the proposed Thirty Meter Telescope. By late 2006, an ESO study group had drawn up

**Cerro Armazones
nighttime panorama**
This panorama shows
Cerro Armazones in
the Chilean desert,
the site for the E-ELT
— the world's big-
gest eye on the sky.

a reference design for what would become the European Extremely Large Telescope, and the ESO Council allocated 67 million euros to carry out a design study.

With two potential competitors on the horizon — the GMT and the TMT — Europe realised it had to act decisively to retain its leading position in ground-based optical and infrared astronomy. Thanks to a technological design study sponsored by the European Commission and a convincing science case, the E-ELT ended up prominently in the roadmap of the European Strategy Forum on Research Infrastructure, as well as in the ASTRONET European Infrastructure Roadmap for Astronomy. Within a few years, the project had gained momentum and visibility.

But what about the cost? With an estimated construction budget of over one billion euros, the E-ELT would require additional funding from ESO's Member States equal to three annual contributions over the next decade. Even

then, an additional injection would be needed. Luckily, by 2011 Brazil had expressed an interest in joining ESO as the 15th (and first non-European) Member State. Brazil's entrance fee and annual contributions would facilitate the development of the E-ELT. Around the same time the future monster telescope was scaled down a notch to 39.3 metres, to minimise the risk of the project not being finished on time. With Brazil's membership progressing, ESO's Council approved the programme in 2012.

The current design of the E-ELT involves 798 actively controlled hexagonal segments, each 1.4 metres in diameter and only 5 centimetres thick. The secondary and tertiary mirrors will measure 4.2 and 3.7 metres across, respectively. Two additional mirrors provide the telescope with adaptive optics, through the use of no less than four laser guide stars and over a thousand computer-controlled actuators. All in all, the E-ELT will collect a hundred million times more light than the human eye; it will be 15 times

E-ELT at sunset
The E-ELT will be a revolutionary new ground-based telescope and will have a 40-metre-class main mirror.

Assembled E-ELT mirror segments undergoing testing
Four segments of the giant primary mirror of the E-ELT undergoing testing together for the first time. The assembly, at ESO's facility in Germany, provides a full-size mock-up of a small section of the E-ELT primary mirror and its support structures.

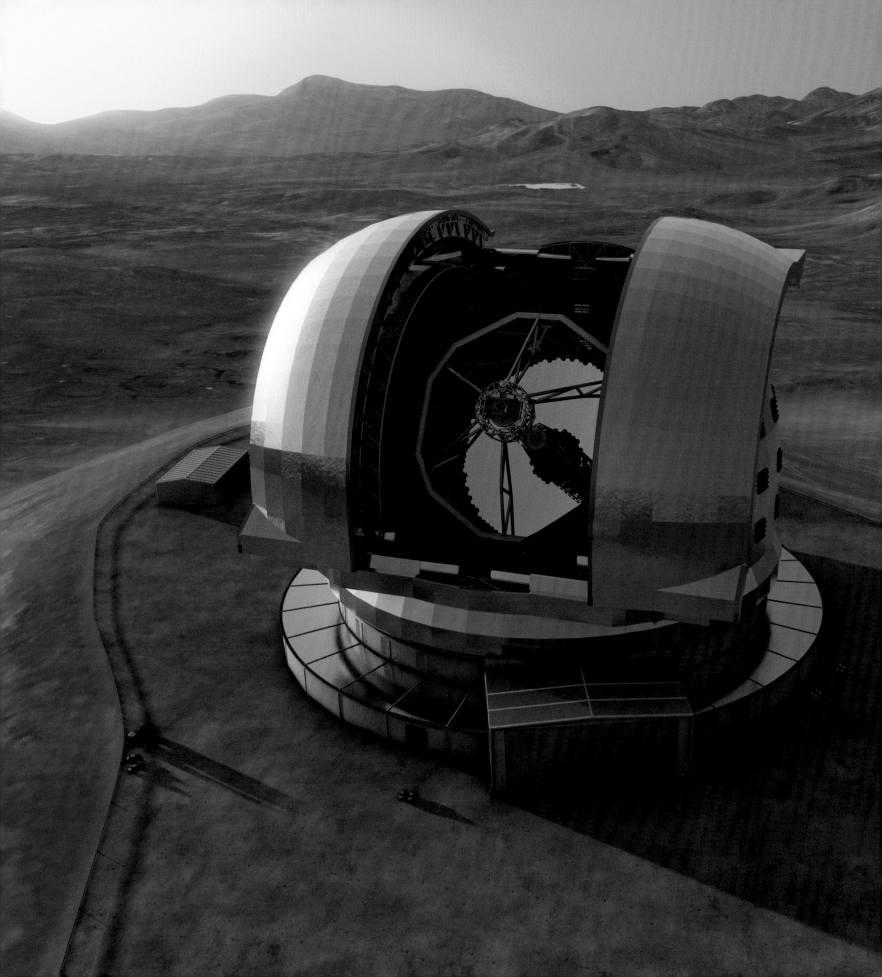

as sensitive as the current generation of 8- to 10-metre-class telescopes, and it will see 15 times more detail than the Hubble Space Telescope.

The novel design for the E-ELT's mount is based on a paradigm shift in telescope building: a modular, Lego-like design with many identical parts that can be mass-produced to save costs. The telescope structure weighs in at no less than 2700 tonnes, yet it can be precisely aimed at any position in the sky in just a few minutes.

Two 150-tonne platforms as big as tennis courts, one at each end of the telescope's horizontal axis, will house the huge cameras and spectrographs that constitute the electronic retina of ESO's giant eye on the sky. A hemispherical dome, 86 metres in diameter and 74 metres tall, protects the telescope from the harsh desert environment. At night, two giant sliding doors with a total width of over 45 metres will open up to provide the 39.3-metre mirror with an unobstructed view of the Universe.

The future of exoplanet research

The first extrasolar planet orbiting a Sun-like star was discovered by Swiss astronomers Michel Mayor and Didier Queloz in 1995. Since then, hundreds of exoplanets have been found, and there are a few thousand candidate planets awaiting confirmation. Currently, NASA's Kepler space telescope is one of the most prolific exoplanet hunters, looking for the minute periodic brightness dips in the light of stars that occur when orbiting planets pass in front of them. Meanwhile, Mayor's team is using the HARPS spectrograph at ESO's 3.6-metre telescope at La Silla to detect small periodic wobbles of stars that are produced by the gravitational tugs of orbiting planets.

The future of exoplanet research will see three important developments. First, the Kepler mission, extended through 2016, will almost certainly succeed in finding the first Earth-like planet orbiting in the so-called habitable zone of a Sun-like star — a true Earth analogue. Moreover, Kepler will vastly increase the number of known exoplanetary systems and improve our statistical knowledge about the frequency of Earth-like planets in outer space.

Second, the European Space Agency's Gaia mission, due for launch in 2013, will accurately measure the positions and velocities of about one billion stars in the Milky Way. Many of these stars are expected to show a minute positional variation on the sky, and Gaia's astrometrical measurements will also reveal the existence of hundreds, if not thousands of exoplanets, including planets at larger distances from their parent star.

Third, ground-based observatories, including ESO's Very Large Telescope, the Atacama Large Millimeter/submillimeter Array, and the future European Extremely Large Telescope will scrutinise known nearby exoplanetary systems, with the specific goal of actually imaging the orbiting planets and studying the starlight they reflect. Choosing longer observing wavelengths, like infrared and submillimetre waves, will increase the chances of success, because at these wavelengths, planets are generally brighter and stars are generally dimmer than at optical wavelengths, decreasing the enormous contrast between the bright parent stars and their almost indistinguishable planets. Measuring the amount of polarised light will also improve the detection efficiency: starlight is almost unpolarised, while the light reflected by a planet's surface, ocean or atmosphere is much more strongly polarised.

The Holy Grail of exoplanet research is studying the spectrum of an alien world in detail. That could reveal the existence of bio-markers like oxygen, ozone and methane in the planet's atmosphere. Unfortunately, this is beyond the capabilities of ESO's current suite of astronomical instruments. Astronomers hope that ALMA or the E-ELT will take up the challenge.

Exoplanet HD 85512b
Artists's impression of rocky super-Earth HD 85512b in the southern constellation of Vela (The Sails).

In principle, such studies could reveal the presence of biological activity on an alien world

As for instrumentation, it looks like the E-ELT will start operating early in the 2020s with a sensitive near-infrared camera and an integral field spectrograph — a versatile device that makes it possible to extract spectral information from every point in its field of view at once. At a later stage, four more instruments will be added: a mid-infrared camera and spectrograph, a multi-object spectrograph to precisely study the light of many remote galaxies at once, a high-resolution spectrograph for visible and near-infrared light, and a dedicated planetary camera to image extrasolar planets.

With its unprecedented sensitivity, the European Extremely Large Telescope should be able to detect the very first galaxies that emerged from the cosmic Dark Ages, just a few hundred million years after the Big Bang. At very large distances, astronomers look back to very early stages of cosmic evolution, and observing the birth of these primordial galactic building blocks should reveal details about the origin of the large-scale structure of the Universe. E-ELT measurements might even reveal whether or not the constants of nature have really been constant over billions of years, and shed light on the true nature of dark energy — the mysterious property of empty space that is currently accelerating the expansion of the Universe.

Slightly closer to home, the E-ELT is expected to study individual stars in other galaxies than the Milky Way. Thus, the star formation history and the resulting stellar populations of other "star cities" can be studied. Knowing only about the stars in the Milky Way is like living in Paris and knowing only your nearest neighbours. Learning about stars in other galaxies is like getting to know the demographics and sociology of the population of London, or Tokyo for that matter. The E-ELT might finally solve the remaining riddles surrounding the supermassive black holes in the cores of other galaxies, and reveal how they grew in tandem with the galaxies themselves.

The lifecycles of stars and planets are other topics that the new telescope will help to unravel. With its ultra-sharp vision and its dust-penetrating infrared capabilities, the E-ELT will complement the observations of ALMA, and reveal the detailed characteristics of protostars, protoplanetary discs, bloated red giants and tiny white dwarfs. It will also spot very low-luminosity red dwarf stars; still fainter "failed stars" known as brown dwarfs, and maybe even free-floating giant planets, adding to the inventory of the denizens of the Milky Way. Meanwhile, exploding giant stars like supernovae and gamma-ray bursts will be detected and studied all across the observable Universe.

The hope is that the European Extremely Large Telescope will be sensitive enough to directly image small exoplanets orbiting nearby stars. That would provide astronomers with their first opportunity of dissecting the light of a distant sun after it has been reflected by the surface or the atmosphere of an Earth-like planet. In principle, such studies could reveal the presence of biological activity on an alien world — a revolutionary breakthrough that might actually be much harder to achieve than some optimists are hoping. But even if the E-ELT is not up to the task, there's no doubt that the new telescope will not disappoint its users. After all, every new quantum jump in the development of astronomical telescopes and instrumentation has brought completely unexpected surprises, and the E-ELT will not be an exception to that rule.

Eager to find the best possible location for the new monster telescope, ESO carried out an extensive site-testing campaign at various sites in Chile, Argentina, Morocco, and the Canary Islands. This led to a shortlist of preferred sites, including the Roque de los Muchachos Observatory at La Palma, and four mountaintops in northern Chile: Ventarrones, Tolonchar, Vizcachas, and Armazones, which, incidentally, had also been considered for the American Thirty Meter Telescope. In the end, Armazones turned out to offer the best possible combination of dry, steady and clear skies, with the added bonus of good accessibility: the mountain is located just 20 kilometres from Paranal, which the new telescope will share a lot of infrastructure with.

Director General Tim de Zeeuw

Name: Prof. Pieter Timotheus (Tim) de Zeeuw
Year of Birth: 1956
Nationality: Dutch
Period as Director General: Since September 2007

What makes ESO special?

ESO is special in astronomy as it is an intergovernmental organisation based on a treaty (Convention) between the Member States. This provides long-term stability and planning ability, and allows the Organisation to access the highest levels of government, both in the Member States and in Chile where the telescopes are, and so foster support of fundamental research. It has also led to pre-eminence in ground-based astronomy.

What is the greatest challenge during your current term as ESO's Director General?

One of the greatest challenges in the recent past has been to design the world-leading European Extremely Large Telescope (E-ELT) and secure the extra funding needed to construct it on a competitive time scale. This required internal restructuring as well as fundraising in an economically difficult climate. The pending accession of Brazil provides a further major strategic step for ESO, expanding its membership beyond the boundaries of Europe.

What is your favourite ESO anecdote?

Five weeks into my position as Director General I received an email message containing a large attachment. Upon opening it, the PDF slowly scrolled over my screen, revealing first the famous 007 logo, and then the text of a letter, addressed to me, asking for permission to shoot some scenes of Bond22 (*Quantum of Solace*) on Paranal. My immediate reaction was that this was a hoax perpetrated by one of my "friends", expecting that I would fall for it, being fresh on the new job. Further scrutiny of the cc's on the address list then convinced me that this might actually be real, as it was. And after a number of meetings the scenes were indeed taken. For me it was very gratifying to see that the entire film crew was fascinated by the location of Paranal, and that they were very eager for a special tour of the telescopes.

How do you see ESO's future?

ESO's programme for the next 15 years will be dominated by the construction of the E-ELT, while we continue to make Paranal better and capitalise on the tremendous science potential of ALMA. The number of Member States will increase gradually in line with a strategy that pursues excellence in providing world-class facilities for ground-based astronomy. In the longer term, this means constructing and operating further facilities after the E-ELT construction is completed while retiring some of our current telescopes in due course. The pace of technology is such that new instruments can be envisaged that will further push the frontiers of our science. It is hard to imagine that we will know everything our society wants to know about the Universe we live in by the end of the next few decades, and I therefore believe ESO has a very exciting future ahead.

Cerro Paranal and Cerro Armazones
This aerial view beautifully shows the Chilean Atacama Desert around the ESO Paranal Observatory, home to the Very Large Telescope (seen at the bottom), and Cerro Armazones (above middle, left). The volcano Llullaillaco is seen in the distance (top right), 190 kilometres from Paranal.

Right now, the European Extremely Large Telescope is still a distant dream; Cerro Armazones nothing more than a bare mountain with a small weather station and a telecom mast. But ESO's dreams have a tendency to come true. Fifty years ago, it was Jan Oort's and Walter Baade's dream of a European astronomy outpost in the southern hemisphere that came true with the signing of the ESO Convention. In the 1970s and 1980s, both Cerro La Silla and Cerro Paranal were transformed into world-class observatories bustling with scientific activity. Oort and Baade couldn't even have imagined the technological and scientific miracles of ALMA, at Llano de Chajnantor. And ESO's newest dream will hopefully also be realised, by dedicated and visionary scientists and engineers who

Site testing for the E-ELT
The site testing for the E-ELT was extensive. Several mountaintops were tested through year-long campaigns. These included:
1 Roque de los Muchachos Observatory (La Palma, Spain)
2 Armazones (Chile)
3 Ventarrones (Chile)
4 Tolonchar (Chile)
5 Vizcachas (Chile)
6 Cerro Macon (Argentina)
7 Aklim (Morocco).

The E-ELT on Cerro Armazones (artist's impression)
Artist's impression of the European Extremely Large Telescope (E-ELT) on Cerro Armazones, a 3060-metre mountaintop in Chile's Atacama Desert.

Satellite photo of the Paranal and Armazones region
The VLT and the Paranal base camp are located lower left. The sealed B-710 road is going vertically through the left page. Paranal's airstrip is visible in the lower part of the image parallel to B-710. The unsealed road to Armazones in the upper right corner is the white line cross-ing through the image.

strongly believe in the pursuit of knowledge and the power of cooperation.

The European Southern Observatory is fifty years old, but more vital than ever. What new insights — and new mysteries — will ALMA and the E-ELT bring us fifty years from now? What revolutionary astronomical facilities will we be using in 2062 to learn more about the Universe we live in? Only time will tell. One thing is for sure, though: Europe won't cease to be part of the adventure. The fun has only just begun.

Getting to know the E-ELT

The European Extremely Large Telescope will be the biggest optical/infrared telescope in the history of mankind. It catches more light than all current professional telescopes combined, and it will enable astronomers to study the Universe in unprecedented detail. Developing the E-ELT is not only a major endeavour in science, but also in technology and engineering. Partnerships between institutes, universities and industry have already been formed to build and exploit the telescope and its suite of scientific instruments. In addition, E-ELT-related technologies and innovations have wider applications within industry and in important areas such as medicine. Thus, the E-ELT will have a significant economic, cultural and scientific impact beyond its immediate astronomical significance.

The E-ELT in numbers

- Main mirror: 39.3 metres
- Optical design: a novel five-mirror scheme
- Number of primary mirror segments: 798
- Segment size: 1.45 metres wide and only 50 millimetres thick
- Recoating of segments: two per day, all segments in a year
- Field of view: an area on the sky about one ninth the size of the full Moon
- Budget: 883 million euros + contingency 100 million euros + instruments 100 million euros, total 1083 million euros
- Telescope weight: 2800 tonnes
- Dome height: 73 metres
- Site: Cerro Armazones
- Altitude: 3046 metres

E-ELT mirror support
The support for the E-ELT M4 mirror at the ESO Headquarters in early 2012.

Active Phasing Experiment
Components of the Active Phasing Experiment designed to validate technologies for accurate alignment of the segmented mirror of the forthcoming E-ELT.

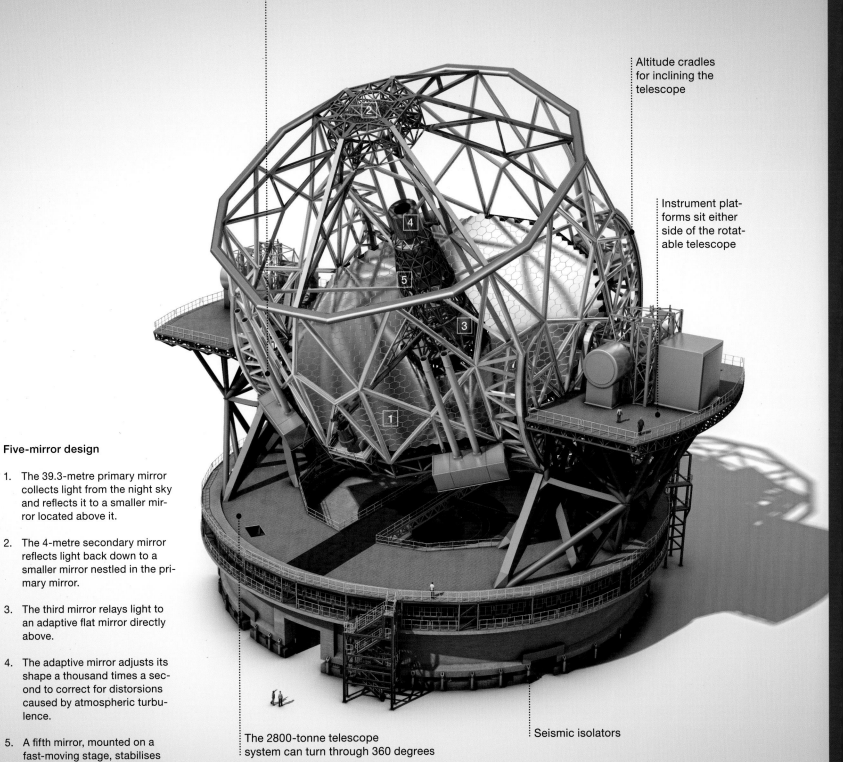

Lasers

Altitude cradles
for inclining the
telescope

Instrument plat-
forms sit either
side of the rotat-
able telescope

Five-mirror design

1. The 39.3-metre primary mirror
    collects light from the night sky
    and reflects it to a smaller mir-
    ror located above it.

2. The 4-metre secondary mirror
    reflects light back down to a
    smaller mirror nestled in the pri-
    mary mirror.

3. The third mirror relays light to
    an adaptive flat mirror directly
    above.

4. The adaptive mirror adjusts its
    shape a thousand times a sec-
    ond to correct for distorsions
    caused by atmospheric turbu-
    lence.

5. A fifth mirror, mounted on a
    fast-moving stage, stabilises
    the image and sends the light to
    cameras and other instruments
    on the stationary platform.

The 2800-tonne telescope
system can turn through 360 degrees

Seismic isolators

Nighttime at Cerro Armazones
Not long from now this peaceful mountain top will be a bustle of activity. Over the period of a decade this summit will be transformed into a world-leading observatory.

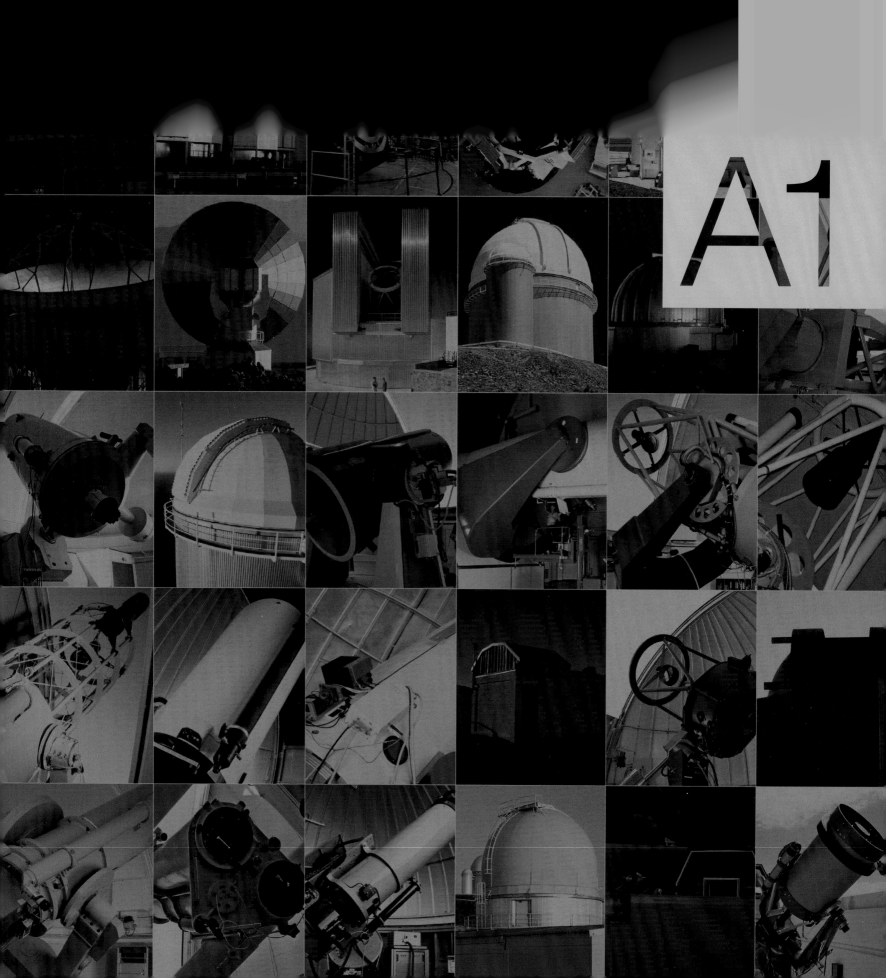

A1

ESO's Telescopes

As set out in its convention, ESO provides state of the art facilities for Europe's astronomers and promotes and organises cooperation in astronomical research. Today, ESO operates some of the world's largest and most advanced observational facilities at three sites in Northern Chile: La Silla, Paranal and Chajnantor. These are the best locations known in the southern hemisphere for astronomical observations. Below is an overview of the 30 different smaller or larger telescope installations that are, or have been, part of ESO.

E-ELT

Dubbed E-ELT for European Extremely Large Telescope, this revolutionary ground-based telescope will have a 40-metre-class main mirror and will be the largest optical/near-infrared telescope in the world: "the world's biggest eye on the sky". It will be operated as part of the Paranal Observatory.

ALMA

The Atacama Large Millimeter/submillimeter Array, or ALMA, is an international collaboration finishing a telescope of revolutionary design to study the Universe from a site in the foothills of Chile's Andes Mountains. ALMA is composed of 66 high precision antennas, operating at wavelengths of 0.32 to 3.6 millimetres.

APEX

APEX, the Atacama Pathfinder Experiment, is a collaboration between the Max-Planck-Institut für Radioastronomie (MPIfR) at 50%, Onsala Space Observatory (OSO) at 23%, and the European Southern Observatory (ESO) at 27% to construct and operate a modified ALMA prototype antenna as a single dish on the 5100 metre high site of Llano Chajnantor.

Paranal Observatory

The Very Large Telescope (VLT) at Cerro Paranal is ESO's premier site for observations in the visible and infrared light. All four Unit Telescopes of 8.2-metre diameter are individually in operation with a large collection of instruments.

La Silla Observatory

ESO's historical site, where more than 20 telescopes have been built over the past 40 years. Today, ESO operates three major telescopes (3.6-metre telescope, New Technology Telescope, 2.2-metre MPG/ESO telescope) at the La Silla Observatory. They are equipped with state-of-the-art instruments either built completely by ESO or by external consortia, with a substantial contribution by ESO.

Paranal Observatory

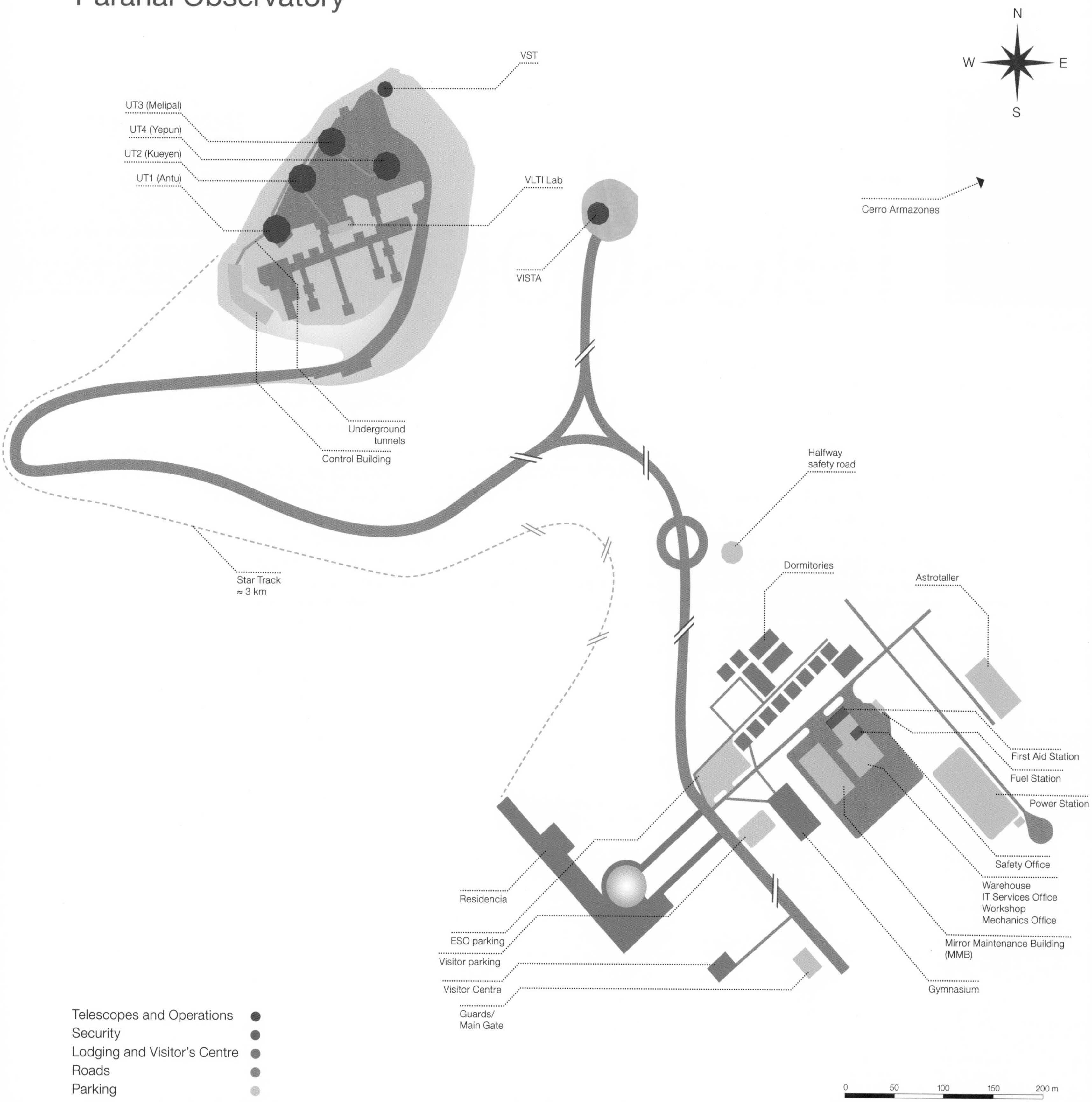

VST

UT3 (Melipal)
UT4 (Yepun)
UT2 (Kueyen)
UT1 (Antu)

VLTI Lab

VISTA

Underground tunnels

Control Building

Star Track ≈ 3 km

Cerro Armazones

Halfway safety road

Dormitories

Astrotaller

First Aid Station

Fuel Station

Power Station

Safety Office

Warehouse
IT Services Office
Workshop
Mechanics Office

Mirror Maintenance Building (MMB)

Gymnasium

Residencia

ESO parking

Visitor parking

Visitor Centre

Guards/
Main Gate

Telescopes and Operations
Security
Lodging and Visitor's Centre
Roads
Parking

N
W — E
S

0   50   100   150   200 m

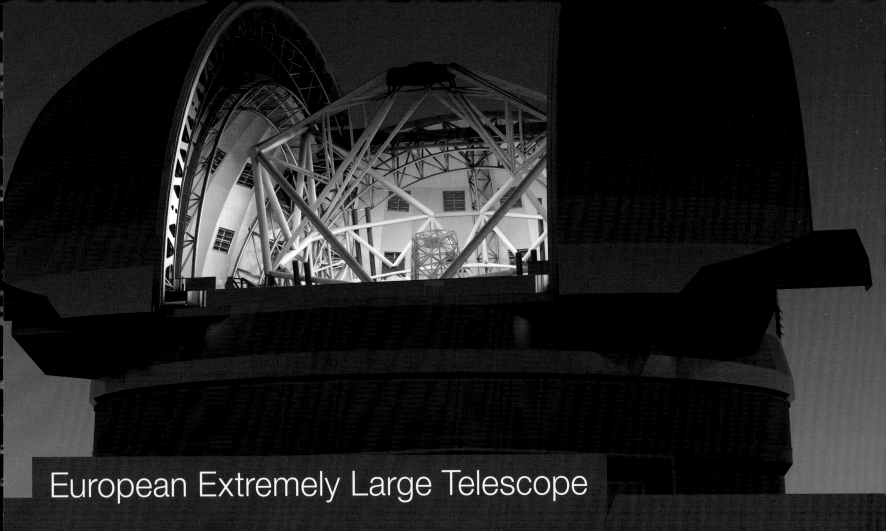

European Extremely Large Telescope

Extremely Large Telescopes are considered worldwide as one of the highest priorities in ground-based astronomy. They will vastly advance astrophysical knowledge, allowing detailed studies of subjects including planets around other stars, the first objects in the Universe, supermassive black holes, and the nature and distribution of the dark matter and dark energy which dominate the Universe.

Dubbed E-ELT for European Extremely Large Telescope, this revolutionary ground-based telescope will have a 40-metre-class main mirror and will be the largest optical/near-infrared telescope in the world: "the world's biggest eye on the sky".

Science goals
General purpose extremely large aperture optical/infrared telescope. Some science areas are to be high redshift galaxies, star formation, exoplanets and protoplanetary systems.

Name:	European Extremely Large Telescope
Site:	Cerro Armazones
Altitude:	3046 m
Enclosure:	Hemispherical dome
Type:	Optical/near-infrared giant segmented mirror telescope
Optical Design:	Five-mirror design: three-mirror on-axis anastigmat + two fold mirrors used for adaptive optics
Diameter. Primary M1:	39.30 m (798 hexagonal 1.4 m mirror segments)
Material. Primary M1:	Not yet decided
Diameter. Secondary M2:	4 m
Material. Secondary M2:	Not yet known
Diameter. Tertiary M3:	3.75 m
Mount:	Alt-azimuth mount
First Light:	Early 2020s
Active Optics:	Yes
Adaptive Optics:	2.60-metre adaptive M4 using four laser guide stars

Very Large Telescope

The Very Large Telescope array is the flagship facility for European ground-based astronomy at the beginning of the third millennium. It is the world's most advanced optical instrument, consisting of four Unit Telescopes with main mirrors of 8.2 metres diameter and four movable 1.8-metre diameter Auxiliary Telescopes. The telescopes can work together, to form a giant interferometer, the ESO Very Large Telescope Interferometer, allowing astronomers to see details up to 25 times finer than with the individual telescopes. The light beams are combined in the VLTI using a complex system of mirrors in underground tunnels where the light paths must be kept equal to distances less than 1/1000 millimetre over a hundred metres. With this kind of precision the VLTI can reconstruct images with an angular resolution of milliarcseconds, equivalent to distinguishing the two headlights of a car at the distance of the Moon.

The 8.2-metre diameter Unit Telescopes can also be used individually. With one such telescope, images of celestial objects as faint as magnitude 30 can be obtained in a one-hour exposure. This corresponds to seeing objects that are four billion (four thousand million) times fainter than those that can be seen with the unaided eye.

Name:	Very Large Telescope
Site:	Cerro Paranal
Altitude:	2635 m
Enclosure:	Compact optimised cylindrical enclosure
Type:	Optical/infrared, with interferometry
Optical Design:	Ritchey-Chrétien reflector
Diameter. Primary M1:	8.20 m
Material. Primary M1:	Zerodur
Diameter. Secondary M2:	1.12 m
Material. Secondary M2:	Beryllium
Diameter. Tertiary M3:	1.24 × 0.87 m (elliptical flat)
Mount:	Alt-azimuth mount
First Light:	UT1, Antu: 25 May 1998
	UT2, Kueyen: 1 March 1999
	UT3, Melipal: 26 Jan 2000
	UT4, Yepun: 4 September 2000
Active Optics:	Yes
Adaptive Optics:	UT4: Laser Guide Star + NACO + SINFONI
Interferometry:	UT maximum 130-metre baseline

Science goals

General purpose large aperture optical/infrared telescope. Applications include high redshift galaxies, star formation, exoplanets and protoplanetary systems.

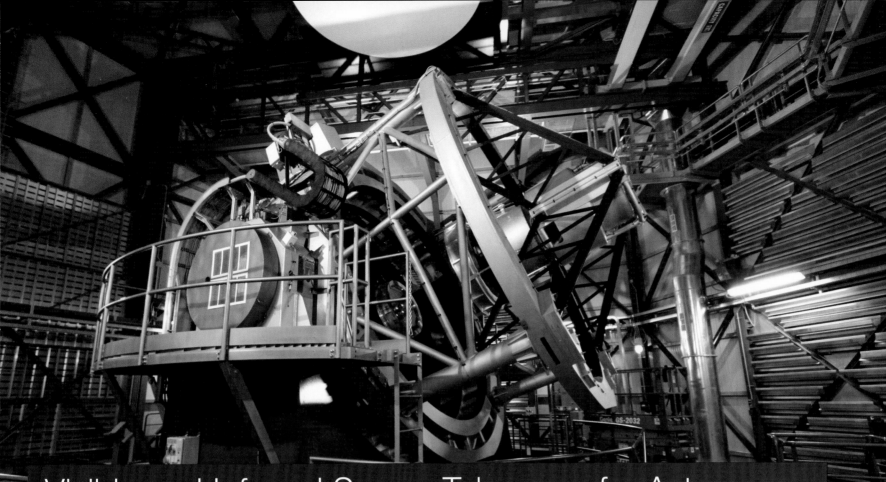

Visible and Infrared Survey Telescope for Astronomy

VISTA — the Visible and Infrared Survey Telescope for Astronomy — is part of ESO's Paranal Observatory. VISTA works at near-infrared wavelengths and is the world's largest survey telescope. Its large mirror, wide field of view and very sensitive detectors will reveal a completely new view of the southern sky.

The telescope is housed on the peak adjacent to the one hosting the ESO Very Large Telescope and shares the same exceptional observing conditions.

VISTA has a main mirror that is 4.1 metres across. In photographic terms it can be thought of as a 67-megapixel digital camera with a 13 000 mm f/3.25 mirror lens.

At the heart of the telescope is a huge three-tonne camera with 16 state-of-the-art infrared-sensitive detectors.

Science goals
Devoted to surveys. Variable stars, deep surveys, brown dwarfs.

Name:	Visible and Infrared Survey Telescope for Astronomy
Site:	Cerro Paranal
Altitude:	2518 m
Enclosure:	Compact optimised cylindrical enclosure
Type:	Near-infrared survey telescope
Optical Design:	Modified Ritchey-Chrétien reflector with corrector lenses in camera
Diameter. Primary M1:	4.10 m
Material. Primary M1:	Zerodur
Diameter. Secondary M2:	1.24 m
Material. Secondary M2:	Astro-Sitall
Mount:	Alt-azimuth fork mount
First Light:	11 December 2009
Active Optics:	Yes

VLT Survey Telescope

The VLT Survey Telescope is the largest telescope in the world designed for surveying the sky in visible light. It is equipped with an enormous 268-megapixel camera called OmegaCAM that is the successor of the very successful Wide Field Imager (WFI) currently installed at the MPG/ESO 2.2-metre telescope on La Silla.

Like the VLT, the new survey telescope will cover a wide-range of wavelengths from ultraviolet through optical to the near-infrared (0.3–1.0 micrometres). But whereas the largest telescopes, such as the VLT, can only study a small part of the sky at any one time, the VST is designed to photograph large areas quickly and deeply.

With a total field view of 1° × 1°, twice as wide as the full Moon, the VST was conceived to support the VLT with wide-angle imaging by detecting and pre-characterising sources, which the VLT Unit Telescopes can then observe further.

Science goals
Devoted to surveys. Remote Solar System bodies such as Trans-Neptunian Objects and Kuiper Belt Objects (TNO, KBO), Milky Way, extragalactic planetary nebulae, cosmology.

Name:	VLT Survey Telescope
Site:	Cerro Paranal
Altitude:	2635 m
Enclosure:	Compact optimised cylindrical enclosure
Type:	Optical survey telescope
Optical Design:	Modified Ritchey-Chrétien
	Reflector with correctors
Diameter. Primary M1:	2.61 m
Material. Primary M1:	Astro-Sitall
Diameter. Secondary M2:	0.94 m
Material. Secondary M2:	Astro-Sitall
Mount:	Alt-azimuth fork mount
First Light:	8 June 2011
Active Optics:	Yes

Auxiliary Telescopes

The four Auxiliary Telescopes are 1.8-metre diameter telescopes that feed light to the Very Large Telescope Interferometer at ESO's Paranal Observatory. Uniquely for telescopes of this size they can be moved from place to place around the VLT platform and are self-contained. They were built by AMOS (Belgium).

The top part of each AT is a round enclosure, made from two sets of three segments, which open and close. Their job is to protect the delicate 1.8-metre telescope from the desert conditions. The enclosure is supported by the boxy transporter section, which also contains electronics cabinets, liquid cooling systems, air-conditioning units, power supplies, and more. During astronomical observations the enclosure and transporter are mechanically isolated from the telescope, to ensure that no vibrations compromise the data collected.

The transporter section runs on tracks, so the ATs can be moved to 30 different observing locations. As the VLT Interferometer (VLTI) acts rather like a single telescope as large as the group of telescopes combined, changing the positions of the ATs means that the VLTI can be adjusted according to the needs

Name:	Auxiliary Telescopes
Site:	Cerro Paranal
Altitude:	2635 m
Enclosure:	Relocatable dome
Type:	Relocatable interferometric telescope
Optical Design:	Ritchey-Chrétien with Coudé optical train
Diameter. Primary M1:	1.82 m
Material. Primary M1:	Zerodur
Diameter. Secondary M2:	0.14 m
Material. Secondary M2:	Zerodur
Diameter. Tertiary M3:	0.16 × 0.11 m (elliptical flat)
Mount:	Alt-azimuth mount
First Light:	AT1: 24 January 2004, AT2: 2 February 2005, AT3: 1 November 2005, AT4: 15 December 2006
Active Optics:	Yes (passive M1, hexapod control of M2)
Adaptive Optics:	in future NAOMI
Interferometry:	UT and AT maximum 202-metre baseline

ALMA

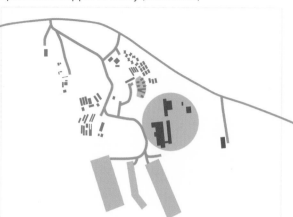

Operations Support Facility (2900m altitude)

Not all of the 28-kilometre route between the Operations Support Facility and the Array Operations Site is shown.

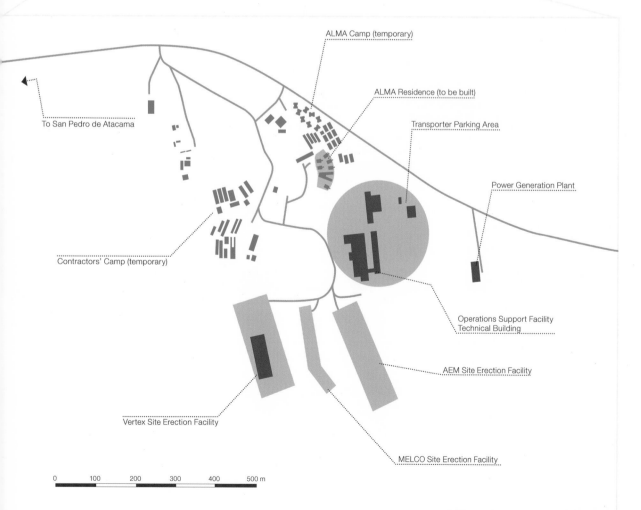

ALMA Camp (temporary)

ALMA Residence (to be built)

Transporter Parking Area

Power Generation Plant

To San Pedro de Atacama

Contractors' Camp (temporary)

Operations Support Facility
Technical Building

AEM Site Erection Facility

Vertex Site Erection Facility

MELCO Site Erection Facility

0 100 200 300 400 500 m

Technical Buildings ●
Lodging and offices ●
Roads ●
Antenna pads ●

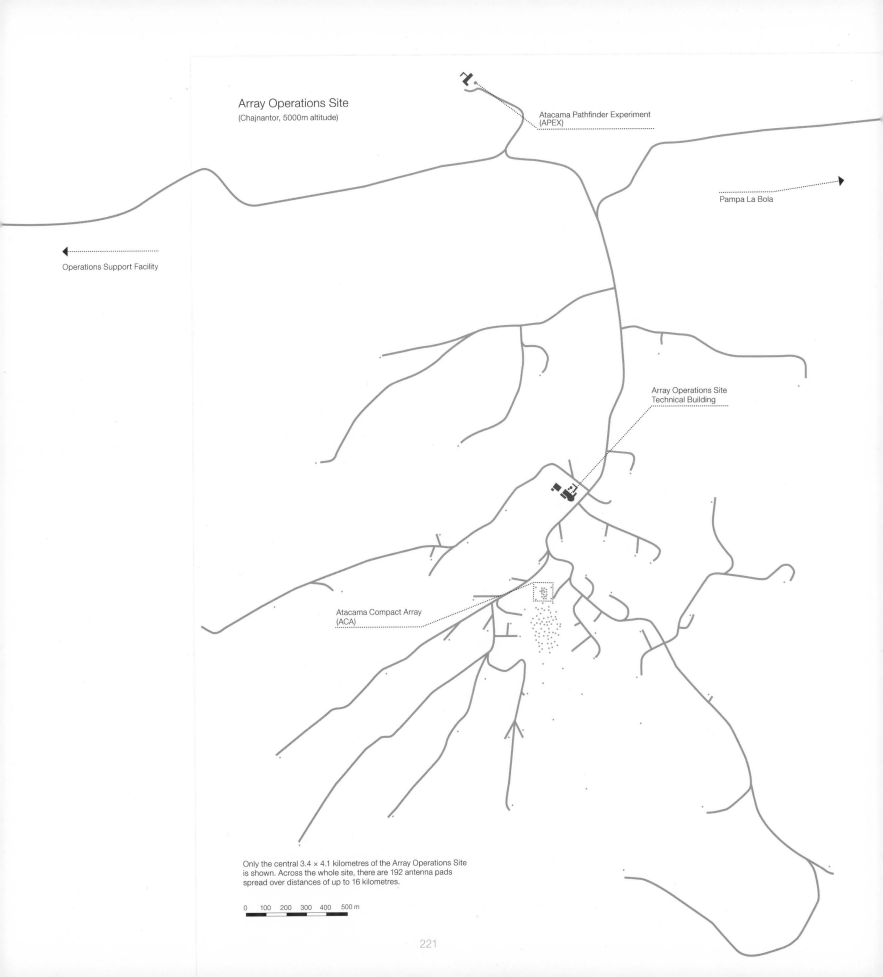

Array Operations Site
(Chajnantor, 5000m altitude)

Atacama Pathfinder Experiment
(APEX)

Pampa La Bola

Operations Support Facility

Array Operations Site
Technical Building

Atacama Compact Array
(ACA)

Only the central 3.4 × 4.1 kilometres of the Array Operations Site
is shown. Across the whole site, there are 192 antenna pads
spread over distances of up to 16 kilometres.

0   100 200 300 400   500 m

Atacama Large Millimeter/submillimeter Array

High on the Chajnantor Plateau in the Chilean Andes, the European Southern Observatory, together with its international partners, is building ALMA — a state-of-the-art telescope to study light from some of the coldest objects in the Universe. This light has wavelengths of around a millimetre, between infrared light and radio waves, and is therefore known as millimetre and submillimetre radiation. ALMA will be composed of 66 high-precision antennas, spread over distances of up to 16 kilometres. This global collaboration is the largest ground-based astronomical project in existence.

Light at these wavelengths comes from vast cold clouds in interstellar space, at temperatures only a few tens of degrees above absolute zero, and from some of the earliest and most distant galaxies in the Universe. Astronomers can use it to study the chemical and physical conditions in molecular clouds — the dense regions of gas and dust where new stars are being born. Often these regions of the Universe are dark and obscured in visible light, but they shine brightly in the millimetre and submillimetre part of the spectrum.

Science goals
Star formation, molecular clouds, early Universe.

Name:	Atacama Large Millimeter/submillimeter Array
Site:	Chajnantor
Altitude:	4576 to 5044m (most above 5000 m)
Enclosure:	Open air
Type:	Submillimetre interferometer antenna array
Optical Design:	Cassegrain
Diameter. Primary M1:	54 × 12.0 m (AEM, Vertex, and MELCO) and 12 × 7.0 m (MELCO)
Material. Primary M1:	CFRP and aluminium (12-metre), Steel and aluminium (7-metre)
Diameter. Secondary M2:	0.75 m (for 12-metre antennas); 0.457 m (for 7-metre antennas)
Material. Secondary M2:	Aluminium
Mount:	Alt-azimuth mount
First Light:	30 September 2011
Interferometry:	Baselines from 150 m to 16 km

Atacama Pathfinder Experiment

ESO operates the Atacama Pathfinder Experiment telescope, APEX, on the Chajnantor Plateau in Chile's Atacama region. APEX is a 12-metre diameter telescope, operating at millimetre and submillimetre wavelengths — between infrared light and radio waves. Submillimetre astronomy opens a window into the cold, dusty and distant Universe, but the faint signals from space are heavily absorbed by water vapour in the Earth's atmosphere.

APEX is the largest submillimetre-wavelength telescope operating in the southern hemisphere. It has a suite of different instruments for astronomers to use in their observations.

APEX is a pathfinder for ALMA, the Atacama Large Millimeter/submillimeter Array, a revolutionary new telescope that ESO, together with its international partners, is now building on the Chajnantor Plateau. APEX is based on a prototype antenna constructed for the ALMA project, and it will find many targets that ALMA will be able to study in great detail.

APEX is a collaboration between the Max Planck Institute for Radio Astronomy (MPIfR, 50%), the Onsala Space Observatory (OSO, 23%), and ESO (27%). The telescope is operated by ESO.

Science goals
Astrochemistry, cold Universe.

Name:	Atacama Pathfinder Experiment
Site:	Chajnantor
Altitude:	5050 m
Enclosure:	Open air
Type:	Submillimetre antenna
Optical Design:	Cassegrain
Diameter. Primary M1:	12.0 m
Material. Primary M1:	CFRP and aluminium
Diameter. Secondary M2:	0.75 m hyperboloidal
Material. Secondary M2:	Aluminium
Mount:	Alt-azimuth mount
First Light:	14 July 2005

La Silla Observatory

Dormitories

Emergency building

Cryogenic plant

Vertex Antenna

Guards/Main Gate ≈ 13 km

Workshop

White House

Heating plant

Clubhouse

Gymnasium

Warehouse

Dormitories

First Aid Station

Bungalow

Hotel

ESO 0.5-metre telescope

Dutch 0.9-metre telescope

Infrastructure &
Support Group (ISG)

New Operations Building (NOB)

ESO 1.52-metre telescope

• (Marly 1-metre telescope)
• (Grand Prisme Objectif telescope)

• TRAnsiting Planets and PlanetesImals
  Small Telescope
• (Swiss T70 telescope)
• (Swiss T40 telescope)

Rapid Eye Mount
telescope (REM)

Danish 1.54-metre telescope

Danish 0.5-metre telescope

ESO 1-metre telescope

MPG/ESO 2.2-metre telescope

Bochum 0.61-metre telescope

Marseille 0.36-metre telescope

Swiss 1.2-metre
Leonhard Euler telescope

ESO 1-metre Schmidt telescope

Workshop/TRS building

Water Tanks & Substation 3

• Télescope à Action
  Rapide pour les Objets
  Transitoires
• (GRB Monitoring System)

DIfferential Motion Monitor (DIMM)

New Technology Telescope (NTT)

Visitor Centre

Coude Auxiliary
Telescope (CAT)

ESO 3.6-metre telescope

Swedish–ESO
Submillimetre Telescope
(SEST), 15-metre

Telescopes and Operations
Technical Buildings
Security
Workshop and offices
Lodging and Visitor Centre
Buildings currently not in use
Roads

N
W E
S

0   100   200   300   400 m

Swedish–ESO Submillimetre Telescope

The Swedish–ESO Submillimetre Telescope was built on behalf of the Swedish Natural Science Research Council and ESO. It was the only large submillimetre telescope in the southern hemisphere at the time of first light. It is very similar to the IRAM telescopes on the Plateau de Bure in France. SEST was decommissioned in 2003, and is superseded by APEX, and ALMA, on Chajnantor.

Science goals
Star formation, molecular clouds.

Name:	Swedish–ESO Submillimetre Telescope
Site:	La Silla
Altitude:	2375 m
Enclosure:	Open air
Type:	Submillimetre antenna
Optical Design:	Cassegrain antennna
Diameter. Primary M1:	15.0 m
Material. Primary M1:	CFRP aluminium
Diameter. Secondary M2:	1.5 m
Mount:	Alt-azimuth mount
First Light:	24 March 1987
Decommissioned:	2003

New Technology Telescope

The 3.58-metre New Technology Telescope was inaugurated in 1989. It broke new ground for telescope engineering and design and was the first in the world to have a computer-controlled main mirror.

The main mirror is flexible and its shape is actively adjusted during observations by actuators to preserve the optimal image quality. The secondary mirror position is also actively controlled in three directions.

This technology, developed by ESO, known as active optics, is now applied to all major modern telescopes, such as the Very Large Telescope at Cerro Paranal and the future European Extremely Large Telescope.

The design of the octagonal enclosure housing the NTT is another technological breakthrough. The telescope dome is relatively small, and is ventilated by a system of flaps that makes air flow smoothly across the mirror, reducing turbulence and leading to sharper images.

Science goals
Star formation, protoplanetary systems, Milky Way centre, spectroscopy.

Name:	New Technology Telescope
Site:	La Silla
Altitude:	2375 m
Enclosure:	Compact optimised enclosure
Type:	Optical and near-infrared telescope
Optical Design:	Ritchey-Chrétien reflector
Diameter. Primary M1:	3.58 m
Material. Primary M1:	Zerodur Schott
Diameter. Secondary M2:	0.88 m
Material. Secondary M2:	Zerodur Schott
Diameter. Tertiary M3:	0.84 m × 0.60 m (elliptical)
Mount:	Alt-azimuth mount
First Light:	23 March 1989
Active Optics:	Yes

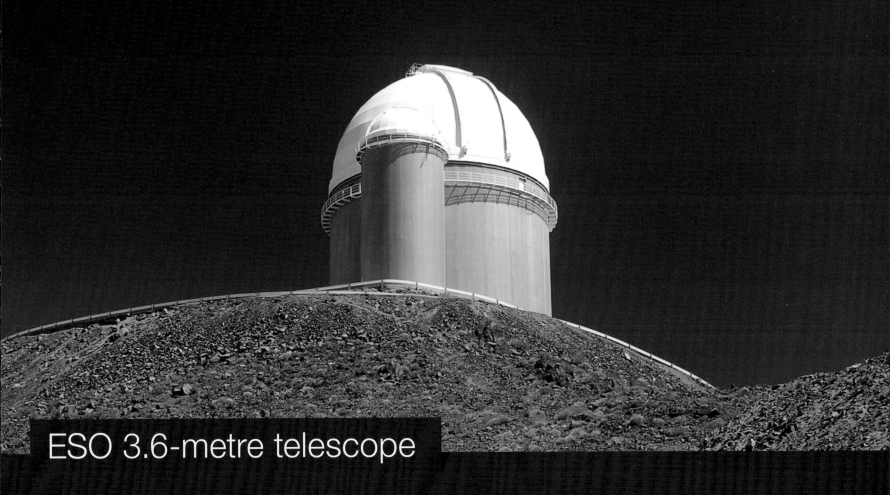

ESO 3.6-metre telescope

The ESO 3.6-metre telescope started operations in 1977 and set Europe the exciting engineering challenge of constructing and operating a telescope in the 3–4-metre class in the southern hemisphere.

Over the years, the ESO 3.6-metre telescope has been constantly upgraded, including the installation of a new secondary mirror that has kept the telescope in its place as one of the most efficient and productive engines of astronomical research.

The telescope hosts HARPS, the High Accuracy Radial velocity Planet Searcher, the world's foremost exoplanet hunter. HARPS is a spectrograph with unrivalled precision and is the most successful finder of low-mass exoplanets to date.

Science goals

Search for exoplanets, asteroseismology.

Name:	ESO 3.6-metre telescope
Site:	La Silla
Altitude:	2375 m
Enclosure:	Classical dome
Type:	Optical and near-infrared telescope
Optical Design:	Cassegrain
Diameter. Primary M1:	3.57 m
Material. Primary M1:	Fused silica
Diameter. Secondary M2:	1.20 m, since Nov 1984: 0.33-metre chopping M2
Material. Secondary M2:	Fused silica
Diameter. Tertiary M3:	1.33 m
Mount:	Equatorial horseshoe mount
First Light:	7 November 1976
Adaptive Optics:	COME-ON, ADONIS 1990

MPG/ESO 2.2-metre telescope

The 2.2-metre telescope has been in operation at La Silla since early 1984 and is on indefinite loan to ESO from the Max Planck Society (Max-Planck-Gesells-chaft or MPG in German). Telescope time is shared between MPG and ESO observing programmes, while the operation and maintenance of the telescope are ESO's responsibility.

The telescope hosts three instruments: the 67-million-pixel Wide Field Imager with a field of view as large as the full Moon, which has taken many amazing images of celestial objects; GROND, the Gamma-Ray Burst Optical/Near-Infrared Detector, which chases the afterglows of the most powerful explosions in the Universe, known as gamma-ray bursts; and the high-resolution spectrograph, FEROS, used to make detailed studies of stars.

Science goals
Gamma-ray burst follow-up, spectroscopy, wide-field imaging, photometry.

Name:	MPG/ESO 2.2-metre telescope
Site:	La Silla
Altitude:	2375 m
Enclosure:	Classical dome
Type:	Optical and near-infrared telescope
Optical Design:	Ritchey-Chrétien reflector
Diameter. Primary M1:	2.20 m
Material. Primary M1:	Zerodur
Diameter. Secondary M2:	0.84 m
Material. Secondary M2:	Zerodur
Mount:	Equatorial fork mount
First Light:	22 June 1983

Danish 1.54-metre telescope

The Danish 1.54-metre telescope, built by Grubb–Parsons, has been in use at La Silla since 1979. It was completely overhauled in 1993 and is now equipped with the Danish Faint Object Spectrograph and Camera spectrograph/camera. The telescope has allowed astronomers to make several first discoveries. In 2005 astronomers showed that short, intense bursts of gamma-ray emission most likely originate from the violent collision of two merging neutron stars, ending a long debate. In 2006, astronomers using a network of telescopes scattered across the globe, including the Danish 1.54-metre telescope, discovered an exoplanet only about five times as massive as the Earth, and circling its parent star in about ten years. This telescope has also been used to produce many impressive astronomical images.

Science goals
Gamma-ray burst follow-up, photometry, microlensing, radial velocities.

Name:	Danish 1.54-metre telescope
Site:	La Silla
Altitude:	2375 m
Enclosure:	Classical dome
Type:	Spectrographic telescope
Optical Design:	Ritchey-Chrétien reflector
Diameter. Primary M1:	1.54 m
Material. Primary M1:	Cervit
Diameter. Secondary M2:	0.61 m
Material. Secondary M2:	Cervit
Mount:	Off-axis equatorial mount
First Light:	20 November 1978

ESO 1.52-metre telescope

The ESO 1.52-metre telescope was essentially a twin of the 1.5-metre telescope at the Observatoire de Haute Provence in France. Two instruments were offered at the ESO 1.52-metre telescope: the B&C spectrograph and FEROS. The time on this telescope was divided between Brazilian national time and ESO time. The telescope was decommissioned at the end of 2002.

Science goals
Stellar spectroscopy.

Name:	ESO 1.52-metre telescope
Site:	La Silla
Altitude:	2375 m
Enclosure:	Classical dome
Type:	Spectrographic telescope
Optical Design:	Cassegrain (f/14.9) or coudé (f/31)
Diameter. Primary M1:	1.52 m
Material. Primary M1:	Borosilicate
Diameter. Secondary M2:	0.43 m (Cassegrain) or 0.36 m (coudé)
Material. Secondary M2:	Borosilicate
Diameter. Tertiary M3:	0.36 m (M3 coudé) and 0.30 m (M4 coudé)
Mount:	English yoke mount
First Light:	7 July 1968
Decommissioned:	Late 2002

Coudé Auxiliary Telescope

The ESO Coudé Auxiliary Telescope (CAT) was housed in a smaller dome, adjacent to the 3.6-metre telescope at ESO's La Silla Observatory and fed the 3.6-metre coudé echelle spectrometer through a light tunnel. The CAT was fully computer controlled and was used for many different types of astronomical research projects, including measuring the ages of ancient stars.

Science goals
High resolution spectroscopy.

Name:	Coudé Auxiliary Telescope
Site:	La Silla
Altitude:	2375 m
Enclosure:	Classical dome connected to ESO 3.6-metre dome
Type:	Spectrographic telescope
Optical Design:	Coudé (feeds the spectrograph of the 3.6-metre)
Diameter. Primary M1:	1.47 m
Material. Primary M1:	Borosilicate
Diameter. Secondary M2:	0.22 m
Material. Secondary M2:	Four interchangeable M2 secondaries mounted on a turret
Diameter. Tertiary M3:	0.254 m
Mount:	Equatorial siderostat mount
First Light:	5 May 1981
Decommissioned:	1 October 1998

Swiss 1.2-metre Leonhard Euler Telescope

The Swiss 1.2-metre Leonhard Euler Telescope at La Silla was built and is operated by the Geneva Observatory (Switzerland) and named in honour of the famous Swiss mathematician Leonhard Euler (1707–83). It is used in conjunction with the CORALIE spectrograph to conduct high precision radial velocity measurements principally to search for large exoplanets in the southern celestial hemisphere. Its first success was the discovery of a planet in orbit around the star Gliese 86. Other observing programmes focus on variable stars, astroseismology, the follow-up of gamma-ray bursts, monitoring of active galactic nuclei and gravitational lenses.

The CORALIE spectrograph, which started operations in June 1998, was developed through a collaboration between the Geneva Observatory and the Haute Provence Observatory (OHP) in France. It is an improved version of the ELODIE spectrograph now in operation at OHP, and with which the first exoplanet was found around the star 51 Pegasi, in 1995. CORALIE is so accurate that it can measure the motion of a star with a precision that is better than 7 m/s or 25 km/h, i.e. about the speed of a fast human runner.

Science goals
Search for exoplanets, asteroseismology, gamma-ray burst follow-up.

Name:	Swiss 1.2-metre Leonhard Euler Telescope
Site:	La Silla
Altitude:	2375 m
Enclosure:	Classical dome
Type:	Optical telescope
Optical Design:	Ritchey-Chrétien reflector
Diameter. Primary M1:	1.20 m
Material. Primary M1:	Zerodur
Diameter. Secondary M2:	0.30 m
Material. Secondary M2:	Zerodur
Diameter. Tertiary M3:	0.24 X 0.17 m (flat elliptical)
Mount:	Alt-azimuth fork mount
First Light:	12 April 1998

ESO 1-metre Schmidt telescope

The ESO 1-metre Schmidt telescope at La Silla began its service life in 1971 using photographic plates to take wide-field images of the southern sky four degrees across — which would cover the full Moon 64 times over. The original photographic camera was decommissioned in December 1998, but the telescope now has a new lease of life as a project telescope. In 2009, a group at Yale's Center for Astronomy and Astrophysics installed a new large camera to conduct a southern hemisphere search for new Pluto-sized dwarf planets and supernovae: the LaSilla–QUEST Variability survey. The camera is a mosaic of 112 CCDs, with a total of 160 million pixels, covering the full field of view of the telescope. The survey is expected to cover about one third of the full sky (about 15 000 square degrees repeated almost every four days). The system is fully operational and controlled remotely from Yale. This project follows the group's northern hemisphere search at Palomar that led to the discovery of the dwarf planet population, including Eris and Sedna.

Science goals
Surveys.

Name:	ESO 1-metre Schmidt telescope
Site:	La Silla
Altitude:	2375 m
Enclosure:	Classical dome
Type:	Astrophotographic telescope
Optical Design:	Schmidt reflector
Diameter. Primary M1:	1.62 m (Schmidt aperture 1.0 m)
Material. Primary M1:	Duran 50 borosilicate
Diameter. Secondary M2:	1.0 m
Material. Secondary M2:	Borosilicate (silicon dioxide)
Mount:	Equatorial fork mount
First Light:	21 December 1971
Decommissioned:	Since 2009 has been carrying out the QUEST survey for Yale University

ESO 1-metre telescope

The ESO 1-metre telescope was the first telescope installed at the La Silla Observatory, in 1966. It was used until 1994 as a photometric telescope, both in the visible with a single channel photometer, and in the infrared with an InSb photometer and a bolometer. Since 1994 it has been fully dedicated to the DENIS project.

Science goals
Stellar photometry, Magellanic clouds, star clusters, stellar associations in the Milky Way.

Name:	ESO 1-metre telescope
Site:	La Silla
Altitude:	2375 m
Enclosure:	Classical dome
Type:	Photometric telescope
Optical Design:	Cassegrain reflector
Diameter. Primary M1:	1.04 m
Material. Primary M1:	Therman low expansion glass (Schott)
Diameter. Secondary M2:	0.295 m (hyperboloid)
Material. Secondary M2:	Fused quartz (Heraeus)
Diameter. Tertiary M3:	0.215 m
Mount:	Equatorial fork mount
First Light:	30 November 1966
Decommissioned:	Since 1994 dedicated to DENIS project

Marly 1-metre telescope

The Marly 1-metre telescope was originally installed in France and operated by groups from Lyon and Marseille, hence the name. It was subsequently moved to La Silla and used for the EROS 2 (Expérience pour la Recherche d'Objets Sombres) search for microlensing events.

Science goals
Microlensing studies, follow-up supernovae.

Name:	Marly 1-metre telescope
Site:	La Silla
Altitude:	2375 m
Enclosure:	Classical dome
Type:	Photometric telescope
Optical Design:	Ritchey-Chrétien reflector
Diameter. Primary M1:	0.98 m
Material. Primary M1:	Zerodur
Diameter. Secondary M2:	0.48 m
Material. Secondary M2:	Zerodur
Mount:	Equatorial fork mount
First Light:	24 June 1996
Decommissioned:	Decommissioned in October 2009 and moved to Mount Djaogari in Burkina Faso

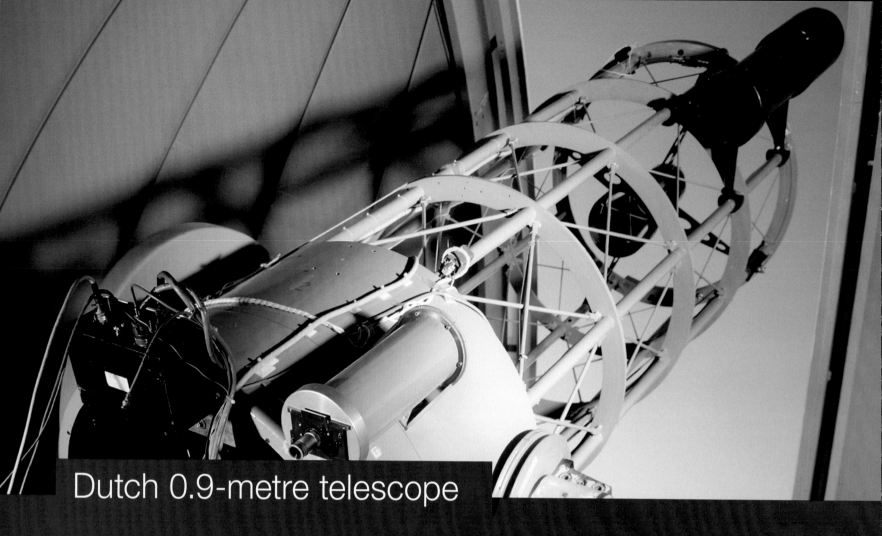

Dutch 0.9-metre telescope

The Dutch telescope was installed at La Silla in 1979 after being at Hartebeespoortdam in South Africa for many years and was originally equipped with a Walraven photometer. It is a reflecting telescope built by Rademakers of Rotterdam, Netherlands in the 1950s. In 1991 the telescope was re-equipped with a CCD and a Cassegrain adapter with two filter wheels and an autoguider. In 2006, the telescope was donated to Alain Maury, who moved it to his observatory near San Pedro de Atacama, where he uses it to search for asteroids.

Science goals
Wide-field imaging, Strömgren narrowband photometry.

Name:	Dutch 0.9-metre telescope
Site:	La Silla
Altitude:	2375 m
Enclosure:	Classical dome
Type:	Astrophotographic telescope
Optical Design:	Dall Kirkham reflector
Diameter. Primary M1:	0.91 m
Material. Primary M1:	Pyrex
Diameter. Secondary M2:	0.25 m
Material. Secondary M2:	Pyrex
Mount:	Equatorial fork mount
First Light:	3 March 1979
Decommissioned:	2006

Swiss T70 telescope

The Swiss T70 telescope was installed on La Silla in 1980, replacing the Swiss 0.4-metre telescope, the previous Swiss national telescope. Like the Swiss 0.4-metre telescope, this telescope was equipped with the P7 photometer, which is a two-channel photometer for quasi-simultaneous measurements in the 7-filter Geneva photometric system. In 1998, when the Swiss 1.2-metre Leonhard Euler Telescope became operational at La Silla, the Swiss T70 telescope was decommissioned.

Science goals
High precision photometry.

Name:	Swiss T70 telescope
Site:	La Silla
Altitude:	2375 m
Enclosure:	Classical dome
Type:	Photometric telescope
Optical Design:	Cassegrain reflector
Diameter. Primary M1:	0.72 m
Material. Primary M1:	Low expansion E3 (Schott)
Diameter. Secondary M2:	0.17 m
Material. Secondary M2:	Fused silica
Mount:	Equatorial fork mount
First Light:	8 August 1980
Decommissioned:	1998

Bochum 0.61-metre telescope

The Bochum 0.61-metre telescope was installed on La Silla in 1968 following a trilateral agreement between ESO, the Deutsche Forschungsgemeinschaft (German Research Foundation) and the University of Bochum. This telescope has the honour of being the first national telescope — belonging to one of the Member States — in ESO's history. The telescope, manufactured by Boller & Chivens, was equipped with a photoelectric photometer made at the central workshop of Göttingen University.

Science goals
Photometry.

Name:	Bochum 0.61-metre telescope
Site:	La Silla
Altitude:	2375 m
Enclosure:	Classical dome
Type:	Photometric telescope
Optical Design:	Cassegrain reflector
Diameter. Primary M1:	0.61 m
Material. Primary M1:	Low expansion silica
Diameter. Secondary M2:	0.15 m
Material. Secondary M2:	Low expansion silica
Mount:	Equatorial cross-axis mount
First Light:	7 September 1968

Rapid Eye Mount telescope

The Rapid Eye Mount telescope is a 60-centimetre rapid-reaction automatic telescope at La Silla, and since October 2002 it has been operated by the REM team for the INAF (Italian National Institute for Astrophysics), a distributed group with its headquarters at the Brera Observatory (Italy). The main purpose of the REM telescope is to follow up promptly the afterglows of gamma-ray bursts detected by the NASA/ASI/STFC Swift satellite. REM is triggered by a signal from Swift or other satellites and quickly points to the designated area. In 2007, thanks to REM, astronomers measured the velocity of the material from the explosions known as gamma-ray bursts for the first time. The material is travelling at an extraordinary speed, more than 99.999% of the velocity of light.

Science goals
Gamma-ray burst follow-up, interstellar and intergalactic medium, distant Universe.

Name:	Rapid Eye Mount telescope
Site:	La Silla
Altitude:	2375 m
Enclosure:	Classical dome
Type:	Robotic optical and near-infrared telescope
Optical Design:	Ritchey-Chrétien reflector
Diameter. Primary M1:	0.60 m
Material. Primary M1:	Astro Sitall
Diameter. Secondary M2:	0.23 m
Material. Secondary M2:	Astro Sitall
Mount:	Alt-azimuth mount
First Light:	25 June 2003

TRAnsiting Planets and PlanetesImals Small Telescope

TRAPPIST (TRAnsiting Planets and PlanetesImals Small Telescope) is a 60-cen-
timetre telescope at La Silla devoted to the study of planetary systems and it
follows two approaches: the detection and characterisation of exoplanets and
the study of comets orbiting around the Sun. The robotic telescope is operated
from a control room in Liège, Belgium. The project is led by the Department
of Astrophysics, Geophysics and Oceanography of the University of Liège, in
close collaboration with the Geneva Observatory (Switzerland). TRAPPIST is
mostly funded by the Belgian Fund for Scientific Research with the participation of the Swiss National Science Foundation.

The name TRAPPIST was given to the telescope to underline the Belgian origin of the project. Trappist beers are famous all around the world and most of
them are Belgian.

Science goals
Search for exoplanets, comets, Trans-Neptunian Objects.

Name:	TRAnsiting Planets and PlanetesImals Small Telescope
Site:	La Silla
Altitude:	2375 m
Enclosure:	Classical dome
Type:	Robotic optical telescope
Optical Design:	Lightweight Ritchey-Chrétien reflector
Diameter. Primary M1:	0.60 m
Material. Primary M1:	Astro Sitall
Diameter. Secondary M2:	0.21 m
Material. Secondary M2:	Astro Sitall
Mount:	German equatorial mount
First Light:	8 June 2010

ESO 0.5-metre telescope

The ESO 0.5-metre telescope was installed on La Silla in 1971. This telescope was a duplicate of the Danish 0.5-metre telescope, and both were manufactured in Copenhagen, Denmark. Initially, it was equipped with a one-channel photometer. It was acquired by ESO in the context of the developments for the control systems of the ESO 3.6-metre telescope, so that these could first be tried out in practice on a small instrument. After 27 years of service, the telescope was decommissioned and moved to the Universidad Católica de Santiago.

Science goals
Photometry.

Name:	ESO 0.5-metre telescope
Site:	La Silla
Altitude:	2375 m
Enclosure:	Classical dome
Type:	Photometric telescope
Optical Design:	Cassegrain reflector
Diameter. Primary M1:	0.52 m
Diameter. Secondary M2:	0.15 m
Mount:	Equatorial fork mount
First Light:	7 December 1971
Decommissioned:	1998. Later moved to Universidad Católica de Santiago

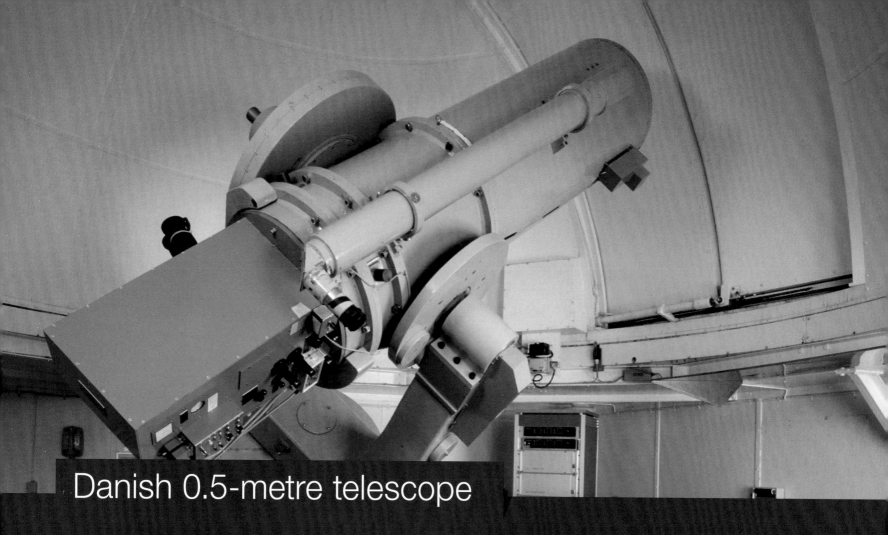

Danish 0.5-metre telescope

The Danish 0.5-metre telescope has been installed on La Silla in 1971, together with its twin, the ESO 0.5-metre telescope. It was equipped with a revolutionary photometer that permitted simultaneous observations in all the filters of the Strömgren system. Moreover, the telescope was fully programmable, which made it a precursor of automatic telescopes.

Science goals
Strömgren narrowband photometry.

Name:	Danish 0.5-metre telescope
Site:	La Silla
Altitude:	2375 m
Enclosure:	Classical dome
Type:	Photometric telescope
Optical Design:	Dall Kirkham Cassegrain reflector
Diamater. Primary M1:	0.50 m
Material. Primary M1:	Low expansion silica
Diameter. Secondary M2:	0.11 m
Material. Secondary M2:	Low expansion silica
Mount:	Equatorial fork mount
First Light:	2 February 1969

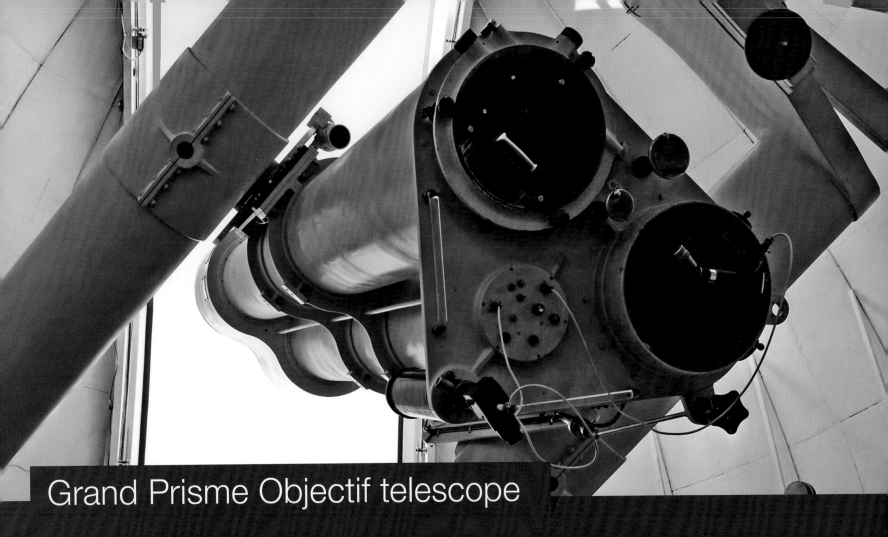

Grand Prisme Objectif telescope

The Grand Prisme Objectif (GPO) is a copy of a telescope installed at the Observatoire de Haute Provence in France. One half of the telescope is a 40-centimetre refractor with an objective prism for obtaining the spectra of many objects at the same time. The other telescope is for guiding. The GPO was first operated in Zeekoegat, South Africa from 1961. The Grand Prisme Objectif was ESO's only refracting (lens) telescope.

Science goals
Astrometry, stellar spectra for classification and radial velocities, minor planet discovery.

Name:	Grand Prisme Objectif telescope
Site:	La Silla
Altitude:	2375 m
Enclosure:	Classical dome
Type:	Double astrograph
Optical Design:	Refractor
Diameter Objective:	0.40 m lens (visual/guiding tube and second photographic tube)
Material Objective:	Doublet Ross lens (flint glass & crown-barium)
Mount:	Equatorial yoke mount
First Light:	6 June 1968
Decommissioned:	In 1996 and replaced by 1-metre Marly telescope.

Swiss T40 telescope

The Swiss T40 telescope was installed at La Silla in 1975, following a convention established in 1974, in which the Council of ESO authorised the Geneva Observatory to set up a provisional observing station on La Silla. This telescope was equipped with a classical photoelectric photometer built in the Geneva Observatory, on which the seven filters of the photometric system of the Geneva Observatory were mounted. The telescope was decommissioned in March 1980, being replaced by the Swiss T70 telescope.

Science goals
Photometry.

Name:	Swiss T40 telescope
Site:	La Silla
Altitude:	2375 m
Enclosure:	Classical dome
Type:	Photometric telescope
Optical Design:	Cassegrain reflector
Diameter. Primary M1:	0.40 m
Mount:	Equatorial mount
First Light:	10 November 1975
Decommissioned:	March 1980 and replaced by Swiss T70 telescope.

Marseille 0.36-metre telescope

The Marseille 0.36-metre telescope was installed on La Silla in September 1989 near the Grand Prisme Objectif telescope, after being assembled and tested at Marseille Observatory in March 1989. The telescope was equipped with a focal reducer, a scanning Fabry-Pérot interferometer and a photon-counting camera. It was devoted to a survey of the Milky Way and the Magellanic Clouds over several years.

Science goals

Fabry-Pérot interferometry of galactic emission nebulae, which is a classical topic at the Observatoire de Marseille (Pérot was from Marseille).

Name:	Marseille 0.36-metre telescope
Site:	La Silla
Altitude:	2375 m
Enclosure:	Sliding roof
Type:	Spectrographic telescope
Optical Design:	Richey-Chrétien reflector
Diameter. Primary M1:	0.36 m
Material. Primary M1:	Low expansion crystallised glass-ceramic
Diameter. Secondary M2:	0.11 m
Material. Secondary M2:	Low expansion crystallised glass-ceramic
Mount:	Equatorial yoke mount
First Light:	20 September 1989

Télescope à Action Rapide pour les Objets Transitoires

The 25-centimetre TAROT (Télescope à Action Rapide pour les Objets Transitoires—Rapid Action Telescope for Transient Objects) is a very fast moving optical robotic telescope on La Silla. It is able to react very quickly to a signal from a satellite indicating that a gamma-ray burst is in progress and can provide fast and accurate positions of transient events within seconds. The data from TAROT will also be useful for studying the evolution of bursts, the physics of the fireball and of the surrounding material. A twin of TAROT is located at the Calern Observatory, in France. Both are operated by a consortium led by Michel Boër (Observatoire de Haute Provence, France).

Name:	Télescope à Action Rapide pour les Objets Transitoires
Site:	La Silla
Altitude:	2375 m
Enclosure:	Double sliding roof
Type:	Robotic optical telescope
Optical Design:	Hyperbolic Newtonian reflector
Diameter. Primary M1:	0.25 m
Material. Primary M1:	Schott N-SK16
Diameter. Secondary M2:	0.14 m
Material. Secondary M2:	Schott N-SK16
Mount:	Equatorial fork mount
First Light:	9 September 2006

GRB Monitoring System

European groups were among the first involved in the hunt for optical and infra-red counterparts of gamma-ray bursts (GRB). The GRB Monitoring System (GMS) was approved in 1982, and its small Celestron telescopes installed in a small building down the hill from the ESO 3.6-metre telescope. The initiator of the project was Holger Pedersen from ESO.

Science goals
Gamma-ray burst optical follow-up.

Name:	GRB Monitoring System
Site:	La Silla
Altitude:	2375 m
Enclosure:	Sliding roof
Type:	Photometric telescope
Optical Design:	Schmidt-Cassegrain reflector
Diameter. Primary M1:	0.20 m
Material. Primary M1:	Annealed Pyrex
Diameter. Secondary M2:	0.07 m
Material. Secondary M2:	Annealed Pyrex
Mount:	Equatorial fork mount
First Light:	1982
Decommissioned:	1986

A2

ESO
Timeline

The year 2012 marks the 50th anniversary of the European Southern Observatory, the foremost intergovernmental astronomy organisation in the world. This very special year provides a great opportunity to look back at ESO's history through 50 highlights, as it celebrates 50 years of reaching new heights in astronomy.

The signing of the ESO Convention in 1962 and the creation of ESO was the culmination of the dream of leading astronomers from five European countries, Belgium, France, Germany, the Netherlands and Sweden: a joint European observatory to be built in the southern hemisphere to give astronomers from Europe access to the magnificent and rich southern sky by the means of a large telescope. The dream resulted in the creation of the La Silla Observatory in Chile, and eventually led to the construction and operation of a fleet of telescopes, with the 3.6-metre telescope as flagship. In the 1980s the New Technology Telescope brought further pioneering advances such as active optics. This prepared the way for the next step: the construction of the world's most advanced visible-light astronomical observatory, the Very Large Telescope at Cerro Paranal.

Today, the original hopes of the five founding members have not only become reality but — as new Member States have joined over the years — ESO has fully taken up the challenge of its mission to design, build and operate the most powerful ground-based observing facilities on the planet. On the Chajnantor Plateau in northern Chile, together with international partners, ESO is developing and operating the biggest ground-based astronomical project in existence, the Atacama Large Millimeter/submillimeter Array. And ESO is preparing to build the world's biggest eye on the sky, the European Extremely Large Telescope. Constantly at the technological forefront, ESO is ready to tackle new and as yet unimaginable territories of scientific discovery.

21 June 1953
A shared European observatory is discussed for the first time by a group of astronomers at Leiden in the Netherlands.

5 October 1962
Founding members Belgium, France, Germany, the Netherlands and Sweden sign the ESO Convention.

7 November 1963
Chile is chosen as the site for the ESO observatory and the *Convenio* (also known as the *Acuerdo*), the agreement between Chile and ESO, is signed.

26 May 1964
The ESO Council selects the mountain Cinchado Nord — later to become La Silla — as the site of its observatory.

30 October 1964
Acquisition of La Silla Mountain and land for the Chilean headquarters in Vitacura.

25 May 1998
First light for the VLT's first Unit Telescope (UT1), Antu.

4 December 1990
Paranal is selected by ESO as the site for the VLT.

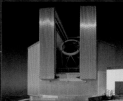

23 March 1989
First light of the New Technology Telescope.

October 1988
The Chilean Government donates the land around Cerro Paranal to ESO.

8 December 1987
Decision is taken by the ESO Council to build the Very Large Telescope.

15 December 1998
Observations of exploding stars, made with telescopes including some at La Silla, show that the expansion of the Universe is accelerating. The 2011 Nobel Prize in Physics was awarded for this result.

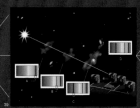

5 March 1999
Official inauguration of Paranal Observatory.

17 March 2001
First light for the Very Large Telescope Interferometer.

5 April 2001
ESO signs an agreement with representatives from North America to build ALMA on the Chajnantor Plateau (Japan joined in 2004).

7 May 2001
Portugal formally joins ESO (Member State 9).

18 November 2008
VLT and APEX studies of violent flares from the centre of the Milky Way reveal material being stretched out as it orbits in the intense gravity close to the central supermassive black hole.

13 May 2008
The VLT detects carbon monoxide in a galaxy seen as it was almost 11 billion years ago, allowing the most precise measurement of the cosmic temperature at such a remote epoch.

30 April 2007
The Czech Republic formally joins ESO (Member State 13).

14 February 2007
Spain formally joins ESO (Member State 12)

11 December 2006
The ESO Council agrees to proceed with studies for the European Extremely Large Telescope.

10 December 2008
ESO's flagship telescopes were used in a 16-year-long study to obtain the most detailed view of the surroundings of the supermassive black hole at the heart of our galaxy.

1 July 2009
Austria formally joins ESO (Member State 14).

11 December 2009
VISTA, the pioneering new survey telescope, starts work.

13 January 2010
The first direct spectrum of an exoplanet is observed with the VLT.

26 April 2010
Cerro Armazones is chosen as the site for the E-ELT.

30 November 1966
First light for the ESO 1-metre telescope at La Silla, the first telescope to be used by ESO in Chile.

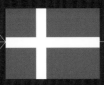

24 August 1967
Denmark formally joins ESO (Member State 6).

25 March 1969
Inauguration of the ESO site at La Silla by the President of the Republic of Chile, Eduardo Frei Montalva, and of ESO's Chilean headquarters in Santiago's Vitacura district.

2 December 1975
The ESO Council approves Garching bei München, Germany, as the new home for ESO's Headquarters.

7 November 1976
First light for the ESO 3.6-metre telescope.

22 June 1983
First light for the MPG/ESO 2.2-metre telescope.

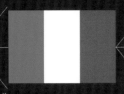

24 May 1982
Italy formally joins ESO (Member State 8).

1 March 1982
Switzerland formally joins ESO (Member State 7).

5 May 1981
Inauguration of the new ESO Headquarters in Garching, Germany.

1978
Completion of the Quick Blue Survey done with the ESO 1-metre Schmidt telescope.

24 June 2002
The United Kingdom formally joins ESO (Member State 10).

11 February 2003
First light of the High Accuracy Radial Velocity Planet Searcher (HARPS) at ESO's 3.6-metre telescope at the La Silla Observatory.

25 July 2003
The Republic of Chile grants free concession of the land on Chajnantor for the ALMA project.

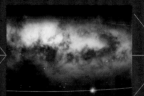

6 April 2004
After 15 years and more than 1000 nights of observations at La Silla, astronomers show from the motions of more than 14 000 Sun-like stars that our galaxy has led a much more turbulent life than previously assumed.

7 July 2004
Finland formally joins ESO (Member State 11).

28 January 2006
First light of the VLT laser guide star, on the VLT's UT4, Yepun.

6 October 2005
ESO telescopes provide definitive proof that long gamma-ray bursts are linked with the ultimate explosions of massive stars, solving a long-standing puzzle.

14 July 2005
First light for the submillimetre Atacama Pathfinder Experiment.

10 September 2004
The VLT obtains the first-ever image of a planet outside the Solar System.

17 August 2004
Using the VLT, astronomers measure the age of the oldest star known in the Milky Way: 13.2 billion years old.

24 August 2010
Astronomers using HARPS discover the richest planetary system so far, containing at least five planets around the Sun-like star HD 10180.

29 December 2010
Brazil signs the Accession Agreement to become a member of ESO.

8 June 2011
First images from the VLT Survey Telescope.

30 September 2011
ALMA starts Early Science and the first image is published.

11 June 2012
The ESO Council approves the European Extremely Large Telescope Programme.

Image Credits

Cover, The VLT: ESO/B. Tafreshi (twanight.org)

Inside front cover, The giant globular cluster Omega Centauri: ESO/INAF-VST/OmegaCAM. Acknowledgement: A. Grado/INAF-Capodimonte Observatory

Back, Heavenly wonders: ESO

p. 6, The VLT at work: ESO/B. Tafreshi (twanight.org)

p. 8, The Chilean night sky at ALMA: ESO/B. Tafreshi (twanight.org)

p. 10, The southern sky at the coast of the Chilean Atacama Desert: G. Hüdepohl (www.atacamaphoto.com)/ ESO

p. 12, Jan Oort: Leiden Observatory

p. 12, The Small Magellanic Cloud over the Chilean landscape: G. Hüdepohl (www.atacamaphoto.com)/ESO

p. 14–15, A 340-million-pixel Paranal starscape: ESO/S. Guisard (www.eso.org/~sguisard)

p. 16, The Southern Cross: ESO/ Y. Beletsky

p. 17, *Terra incognita* of the heavens: images courtesy of Daniel Crouch Rare Books (www.crouchrarebooks.com)

p. 18, The Royal Observatory at the Cape of Good Hope: Chris de Coning/ South African Library/Warner-Madear

p. 19, Director General Otto Heckmann: ESO

p. 20, Star trails over the site-testing station in South Africa: ESO/ W. Schlosser

p. 22, Birth of ESO: ESO/A. Blaauw

p. 23, Jan Oort: Leiden Observatory

p. 24, Site-testing station in South Africa: ESO/W. Schlosser

p. 25, Site-testing in the Karoo, South Africa: J. Dommaget/ESO and J. Boulon/ESO

p. 26, On horseback to Cerro Morado: ESO/F. K. Edmondson

p. 26, Building the ESO 1-metre telescope: ESO

p. 27, Adriaan Blaauw: ESO

p. 28, Map of the northern part of Chile: ESO

p. 29, An early La Silla: ESO/ J. Dommaget

p. 30, La Silla soon after sunset: ESO/José Francisco Salgado (josefrancisco.org)

p. 32, Supernova 1987A in the Large Magellanic Cloud: ESO

p. 33, Star trails over La Silla: ESO/ J. Pérez

p. 34–35, Aerial view of La Silla: ESO

p. 36, The MPG/ESO 2.2-metre telescope: ESO/José Francisco Salgado (josefrancisco.org)

p. 37, Star trails over the ESO 3.6-metre telescope: ESO/A. Santerne

p. 38, The Rapid Eye Mount telescope: ESO

p. 38, TAROT: ESO

p. 39, TRAPPIST: E. Jehin/ESO

p. 40–42, Nighttime at La Silla in 2011: ESO/José Francisco Salgado (jose-francisco.org)

p. 43–46, A panorama of a unique cloudscape over La Silla: ESO/ F. Kamphues

p. 47–49, The ridge of La Silla: ESO/José Francisco Salgado (josefrancisco.org)

p. 50, The ESO 3.6-metre telescope at La Silla: Iztok Bončina/ESO

p. 51, SEST at La Silla: Iztok Bončina/ ESO

p. 51, Comet Shoemaker-Levy 9: ESO

p. 52–53, The Lagoon Nebula: ESO

p. 54, The New Technology Telescope: Iztok Bončina/ESO

p. 55, Lodewijk Woltjer Lecture at JENAM 2010: European Astronomical Society

p. 56, The exoplanet Beta Pictoris b: ESO/L. Calçada

p. 57, The NTT in its enclosure: ESO/C. Madsen

p. 58–59, A full view of the La Silla mountain: ESO/José Francisco Salgado (josefrancisco.org)

p. 60, The Cat's Paw Nebula seen in the infrared with VISTA: ESO/ J. Emerson/VISTA Acknowledgement: Cambridge Astronomical Survey Unit

p. 62, Bok Globule Barnard 68: ESO

p. 62, Colliding galaxies seen with the VLT: ESO

p. 63, Spiral galaxy NGC 1232: ESO

p. 64–65, The Carina Nebula: ESO

p. 66, An artist's rendering of a distant quasar: ESO/M. Kornmesser

Index

Further Reading: Other Books About ESO

Several books have been written about ESO over the years. Here follows a list of some of the most prominent in chronological order.

Sterne, Kosmos, Weltmodelle: Erlebte Astronomie, O. Heckmann, 1976, 360 pages (R. Piper Verlag)
The first ESO Director General, Otto Heckmann, started a wave of books about ESO with his autobiography in German. The last part of the book is dedicated to a historical account of ESO and describes his personal experience starting up a major international scientific organisation in a desert more than 10 000 kilometres from Europe. The challenges were many and the devised solutions often unconventional. This book can at times be found second-hand online.

Exploring the Southern Sky — A Pictorial Atlas from the European Southern Observatory (ESO), S. Laustsen, C. Madsen and R. M. West, 1987, 276 pages (Springer-Verlag: Berlin, Heidelberg)
This comprehensive pictorial atlas from the European Southern Observatory has made views of the southern skies available to many an armchair astronomer and today gives a good baseline reference for the state of astronomical photography 25 years ago. The book is available in several languages including English, German, Spanish, Danish, and French. A full PDF version of the English edition of this large-format picture book can be downloaded for free from: http://www.eso.org/public/products/books/exploring_the_southern_sky/

ESO's Early History — The European Southern Observatory from Concept to Reality, A. Blaauw, 1991, 270 pages (ESO, Garching bei München)
Former ESO Director General Adriaan Blaauw's classical book about ESO's early years is a reference for anyone thirsting for detail. Blaauw was closely associated with ESO through most of his life and here passes a rich trove of knowledge on to future generations. A full PDF file can be downloaded for free from: http://www.eso.org/public/products/books/eso_early_history/

Geheimnisvolles Universum — Europas Astronomen entschleirn das Weltall, D. H. Lorenzen, 2002, 208 pages (Kosmos Verlag)
Dirk Lorenzen's richly illustrated book in German was written for ESO's 40th anniversary in 2002. It is a beautiful large-format book that takes the reader on a thorough journey through hand-picked ESO themes from history, technology and science. A full PDF file can be downloaded for free from: http://www.eso.org/public/products/books/geheimnisvolles-universum/

Europe's Quest for the Universe, L. Woltjer, 2006, 328 pages (EDP Sciences, Paris)
Former ESO Director General Lodewijk Woltjer has written an insightful book about the broad scientific and political landscape of European astronomy on the ground and in space from radio, infrared, and visible wavelengths to X-rays, gamma rays and cosmic rays. The book is a tour-de-force that focuses on the roles of ESO and the European Space Agency (ESA), but also national initiatives and interests are touched upon. It can be read by a wide audience: astronomers and space scientists, students, politicians involved in science funding, amateur astronomers and the educated public with some interest in the European science and technology.

Secrets of the Hoary Deep: A Personal History of Modern Astronomy, R. Giacconi, 2008, 432 pages (The Johns Hopkins University Press)
Nobel laureate and former ESO Director General Riccardo Giacconi wrote this book not only as an autobiography but as a history of contemporary astronomy illustrated by his own experiences. Most of the book focuses on the ground-breaking work he and colleagues did in the early days of X-ray astronomy. This was before the field was really born and when most astronomers did not really believe there was much to observe in X-rays. Three important chapters describe how the VLT project was brought to a successful completion, despite significant challenges. The book reveals the science, people, and institutional settings in various astronomical organisations

with emphasis on the technology developments and the management of big projects. The back flap sets the tone of the writing: "Giacconi's story will captivate, inspire, and, at times, possibly infuriate professional and amateur astronomers". Lots of insights to be gained and an entertaining read.

Eyes on the Skies: 400 Years of Telescopic Discovery, G. Schilling & L.L. Christensen, 2009, 132 pages (Wiley-VCH, Weinheim)

Adopted as the official book of the International Year of Astronomy 2009, in which ESO played a leading role, this illustrated history of telescopic discovery spans the range from the first telescopes via the Hubble Space Telescope and NTT, to the E-ELT. The book and its accompanying movie explore the many facets of the telescope — the historical development, the scientific importance, the technological breakthroughs, and also the people behind this ground-breaking invention, their triumphs and failures. The book is available in several languages including English, German, Finnish, Korean, Japanese, Slovenian and Chinese.

The Jewel on the Mountaintop — The European Southern Observatory through Fifty Years, C. Madsen, 2012, 576 pages (Wiley-VCH, Weinheim)

Authored by ESO Senior Advisor Claus Madsen, this book comprises 576 action-packed pages of ESO history and dramatic stories about the people behind the organisation. This is the ultimate historical account of ESO, which tells the story not only of its telescopes in the southern hemisphere, but also about a truly remarkable European success story in research. Ranging from ESO's first telescopes to the future facilities of the next generation, it shows how the improvement of telescope technology leads to a continuously evolving view of the Universe. Produced especially for ESO's 50th anniversary.

ESO's website, www.eso.org

Also ESO's website and social media is a rich source of information about ESO, its telescopes and the science done with them. To just mention a few of the most important links:

- About ESO: http://www.eso.org/public/about-eso.html
- *ESO Messengers* and old ESO Bulletins can be found in the Periodicals section of the ESO Products: http://www.eso.org/public/outreach/products/
- About the telescopes: http://www.eso.org/public/teles-instr/index.html
- ESO top-10 science: http://www.eso.org/public/science/top10.html
- Fifty Years, Fifty Highlights: Timeline with 50 highlights & 50 photos: http://www.eso.org/public/outreach/50years/50highlights50years.html
- Timeline with almost 200 milestones: http://www.eso.org/public/about-eso/timeline.html
- Image archive with 7000 free images: http://www.eso.org/public/images/
- Video archive with 2000 free videos: http://www.eso.org/public/videos/
- Facebook: http://facebook.com/ESOAstronomy
- Twitter: http://twitter.com/ESO
- YouTube: http://youtube.com/ESOObservatory/
- Vimeo: http://vimeo.com/esoastronomy
- Flickr: http://flickr.com/photos/esoastronomy/

About the Movie

To celebrate its 50th anniversary year, ESO has released the documentary *Europe to the Stars — ESO's first 50 years of Exploring the Southern Sky*. The movie, which is attached to the back of this book, captures the story of its epic adventure — a story of cosmic curiosity, courage and perseverance. The story of discovering a Universe of deep mysteries and hidden secrets. The story of designing, building and operating the most powerful ground-based telescopes on the planet.

The movie consists of eight chapters each focusing on an essential aspect of an observatory, while putting things in perspective and offering a broader view on how astronomy is done. From site testing and explaining the best conditions for observing the sky to how telescopes are built and what mysteries of the Universe astronomers are revealing. It has a total duration of 61 minutes. It is produced in full HD (1080p) and is available on blu-ray or DVD. It has a comprehensive bonus section, narration and subtitles in several languages.

- Blu-ray, mastered in full HD
- 61 minutes
- 8 chapters
- Bonus material
- Narration in several languages
- Subtitles in several languages
- Region-free
- Movie and book website: http://www.eso.org/public/outreach/50years/europetothestars.html

Filming at the ESO sites
Part of the film team while filming at Cerro Armazones in Chile. From left: sound engineer Cristian Larrea, host Dr J (Dr Joe Liske), director Lars Lindberg Christensen, producer Herbert Zodet and author Govert Schilling. Not on this picture: art director Martin Kornmesser and 3D animator Luis Calçada.